AMS SHORT COURSE LECTURE NOTES

Introductory Survey Lectures

published as a subseries of

Proceedings of Symposia in Applied Mathematics

Proceedings of Symposia in
APPLIED MATHEMATICS

Volume 66

Applications of Knot Theory

American Mathematical Society
Short Course
January 4–5, 2008
San Diego, California

Dorothy Buck
Erica Flapan
Editors

American Mathematical Society
Providence, Rhode Island

LECTURE NOTES PREPARED FOR THE
AMERICAN MATHEMATICAL SOCIETY SHORT COURSE
APPLICATIONS OF KNOT THEORY
HELD IN SAN DIEGO, CALIFORNIA
JANUARY 4–5, 2008

The AMS Short Course Series is sponsored by the Society's Program Committee for National Meetings. The series is under the direction of the Short Course Subcommittee of the Program Committee for National Meetings.

2000 *Mathematics Subject Classification.* Primary 53A04, 57M25, 57M27, 57R56, 81T45, 82D60, 92C40, 92E10.

Library of Congress Cataloging-in-Publication Data

AMS Short Course Applications of Knot Theory (2008 : San Diego, Calif.)
Applications of knot theory : AMS Short Course Applications of Knot Theory, January 4–5, 2008, San Diego, California / Dorothy Buck, Erica Flapan, editors.
p. cm.—(Proceedings of symposia in applied mathematics ; v. 66)
Includes bibliographical references and index.
ISBN 978-0-8218-4466-3 (alk. paper)
1. Knot theory–Congresses. 2. DNA–Structure–Congresses. I. Buck, Dorothy, 1973– II. Flapan, Erica, 1956– III. Title.
QA612.2.A465 2008
514′.224–dc22 2008044393

♾ The paper used in this book is acid-free and falls within the guidelines
established to ensure permanence and durability.
Visit the AMS home page at `http://www.ams.org/`

10 9 8 7 6 5 4 3 2 1 14 13 12 11 10 09

Contents

Preface

The original motivation for understanding and classifying knots was due to Lord Kelvin who theorized in the 1880s that atoms were knotted or linked vortex rings in the "ether", and that different elements were determined by the knot or link type of the vortex ring. By the early 1900s, Kelvin's theory had been proven wrong. However topologists continued to study the knot theory as an area of pure mathematics. Over the past 20-30 years, knot theory has rekindled its historic ties with biology, chemistry, and physics as a means of creating more sophisticated descriptions of the entanglements and properties of natural phenomena—from strings to organic compounds to DNA.

For example, DNA knots and links have been implicated in a number of cellular processes since their discovery in the late 1960s. In particular, they have been found during replication and recombination, and as the products of protein actions, notably with topoisomerases, recombinases, and transposases. The variety of DNA knots and links observed makes biologically separating and distinguishing these molecules a critical issue. While DNA knots and links can be visualized via electron microscopy, this process can be both difficult and time-consuming. So topological methods of characterizing and predicting their behavior can be helpful.

Chemists have been interested in molecular chirality since Pasteur first described it in 1848. For example, since the two mirror forms of the same molecule can interact with a host's metabolism very differently, predicting whether or not a molecule will be chiral is important to pharmaceutical companies as they develop new medications. While the geometry of a rigid molecule determines whether or not it is chiral, for flexible or even partially flexible molecules, knot theory can play a role in determining chirality.

In addition to the examples described above, there are many other deep interactions between knot theory and various areas of scientific investigation. The 2008 AMS Short Course *Applications of Knot Theory*, on which this volume is based, was intended to introduce the area of applied knot theory to a broad mathematical audience. The aim of the Short Course and this volume, while not covering all aspects of applied knot theory, is to provide the reader with a mathematical appetizer, in order to stimulate the mathematical appetite for further study of this exciting field.

No prior knowledge of topology, biology, chemistry, or physics is assumed. In particular, the first three chapters of this volume introduce the reader to knot theory (by Colin Adams), topological chirality and molecular symmetry (by Erica Flapan), and DNA topology (by Dorothy Buck). The second half of this volume is focused on three particular applications of knot theory. Lou Kauffman presents a chapter on applications of knot theory to physics, Ned Seeman presents a chapter on how topology is used in DNA nanotechnology, and Jon Simon presents a chapter on the statistical and energetic properties of knots and their relation to molecular biology. The articles and their authors are described in more detail below.

Description of articles and their authors

Chapter 1: A Brief Introduction to Knot Theory from the Physical Point of View by Colin Adams

This article introduces the mathematical theory of knots, including Reidemeister moves, surfaces, types of knots, and various invariants associated to knots, including the new superinvariants. It also touches on the stick number for knots and its implications for chemistry.

Colin Adams is the Thomas T. Read Professor of Mathematics at Williams College. He authored the now-standard undergraduate knot theory text, "The Knot Book", and is renowned for his witty and deceptively sophisticated introductory geometry and topology talks. His own research focuses on hyperbolic knots and 3-manifolds. He has involved numerous undergraduates in annual summer research projects at Williams. He is a recipient of the Deborah and Franklin Tepper Haimo Distinguished Teaching Award from the MAA, and has been selected as a Polya Lecturer for the MAA as well as a Sigma Xi Distinguished Lecturer.

Chapter 2: Topological Chirality and Symmetries of Non-rigid Molecules by Erica Flapan

This article explains the concept of chirality and why it is important, and discusses the differences between chemical, geometric, topological, and intrinsic chirality. It then introduces four different techniques to show that a molecule is topologically chiral. The article concludes by presenting different approaches to classifying molecular symmetries. In particular, it compares the point group to the topological symmetry group, and explains how the topological symmetry group can be used to analyze the symmetries of non-rigid molecules.

Erica Flapan is the Lingurn H. Burkhead Professor of Mathematics at Pomona College. Her research is in knot theory, 3-dimensional topology, and applications of topology to chemistry and biology. Her book "When Topology Meets Chemistry", is jointly published by the Mathematical Association of America and Cambridge University Press. From 2000 to 2004, she was the principal investigator on an NSF-CCLI grant entitled "Enhancing the mathematical understanding of students in chemistry". As part of this grant, she developed a course entitled "Problem Solving in the Sciences", to help students with weak math skills succeed in general chemistry, and together with Daniel O'Leary (an organic chemist), she developed an interdisciplinary upper division course entitled "Symmetry and Chirality".

Chapter 3: DNA Topology by Dorothy Buck

This article introduces DNA, and explains the contributions of knot theory to its study. In particular, it explores the topological techniques used to understand both DNA itself, and how it interacts with proteins in the cell. As an extended example, it gives an overview of the tangle model and its variations to understand the molecular process of site-specific recombination. It also discusses various mathematical contributions to several open questions involving DNA, including how a protein can effectively unknot DNA.

Dorothy Buck is a Mathematical Biologist at Imperial College, London in the Department of Mathematics and Centre for Bioinformatics. She specializes in 3-manifold topology and its applications to mathematical biology. Her training is in both mathematics and microbiology. She spent six years working in molecular biology labs at the University of Texas at Austin and Johns Hopkins Medical School. Before joining the faculty at Imperial, she was an NSF Postdoctoral Fellow with Craig Benham at the Genome Center at the University of California at Davis, and an Assistant Professor in the Applied Mathematics Department at Brown University.

Chapter 4: Knots and Physics by Louis H. Kauffman
Knots are mathematical abstractions of the topological properties of rope in physical space. As such, there are immediate relationships of knots with the physics of ropes, weaves, long-chain molecules and other knotting phenomena in nature. There are also beautiful and surprising relationships of knot theory with the structures and methods of statistical mechanics and quantum theory. This article surveys some of the author's favorite interactions between knots and physics.

Louis Kauffman is Professor of Mathematics at University of Illinois-Chicago. He authored the interdisciplinary text "Knots and Physics". He discovered the bracket polynomial state model for the Jones polynomial, and the first direct relationship between statistical mechanics models and knot invariants. As a topologist, he is omnivorous, working in knot theory and its relationships with statistical mechanics, quantum theory, algebra, combinatorics, and more recently, biology. He is the Editor of the Journal of Knot Theory and its Ramifications.

Chapter 5: Synthetic Single-Stranded DNA Topology by Nadrian C. Seeman
The double helical nature of the DNA molecule has a wide variety of topological implications. Most biologists are familiar with the notion that circular DNA molecules are catenanes/links, so that the strands are linked about once every 10 nucleotides. Consequently, biological systems contain topoisomerases which change the linking topology of the molecule, thereby solving a variety of problems in the metabolism of the genetic material. Today, the realm of DNA extends beyond its biological role as a molecule with an unbranched helix axis. Branched DNA molecules exist as intermediates in genetic recombination, but for 25 years synthetic branched DNA molecules have been built for a variety of purposes that are important for nanotechnology and for molecular computation. The ability to assemble branched DNA backbones has enabled the deliberate construction of single-stranded knots, polyhedral catenanes and Borromean rings. New branched DNA motifs have been derived by using techniques from knot theory. Branched DNA molecules have enabled the deliberate construction of periodic and aperiodic DNA crystals. The applications of these systems include analysis of biological systems, nanoelectronics and nanorobotics.

This article presents the features of synthetic DNA topology, from the design of branched, knotted and linked motifs, to the construction of objects, arrays and devices.

Ned Seeman is the Margaret and Herman Sokol Professor of Chemistry at New York University. He founded the field of structural DNA nanotechnology. His lab has characterized the interactions of synthetic DNA knots with topoisomerases, developed a general algorithm for the construction of any DNA knot, synthesized a DNA molecule that can be built to yield four different topological species, and discovered an RNA topoisomerase. For his innovation, he was awarded the Feynman Prize in Nanotechnology, the Emerging Technology Award from Discover Magazine and elected Fellow of the Royal Society of Chemistry. He is the Founding President of the International Society for Nanoscale Science, Computation and Engineering. Most impressively, in 2008 he was awarded the Nichols Medal from the American Chemical Society.

Chapter 6: Long Tangled Filaments by Jonathan Simon

This article considers filaments, from rope and string and hair to DNA and proteins, anything that might be understood as one-dimensional strands wiggling and tangling in three-dimensional space.

If the filaments are short, we can try to describe their exact geometric shape and understand how their shape relates to physical behavior. If the filaments are somewhat long and flexible, then topological knot type can be very useful, as evidenced by the success of topological methods for studying the actions of DNA enzymes. But if the filaments are very long (think of a complicated 3-dimensional scribble) or somehow random (think of a lot of complicated 3-dimensional scribbles) then it may be impractical to try to describe the exact shapes or even knot types. We need to develop a vocabulary of ideas and models that describe physically important geometric/topological properties of long tangled things.

This article presents ideas, experiments, and theorems dealing with packing, curvature, tangling, and knotting of individual complicated filaments as well as statistical ensembles. It explores some of the work that has been done, some open research problems, and some topics that seem well-suited for undergraduate research activities.

Jon Simon is Professor of Mathematics at the University of Iowa. He pioneered the rigorous applications of knot theory to chemistry, in particular by proving the topological chirality of molecular Möbius ladders. He co-developed the idea of Möbius energy of thick knots. His current research also includes particular knotting and tangling of filaments; "energy" of knots; and applications to molecular biology, in particular, knotted DNA loops.

Acknowledgements

We wish to thank the AMS staff, particularly—Wayne Drady, Edward Dunne, Christine Thivierge, and Luann Cole—for their assistance in preparing both the Short Course and this volume.

Dorothy Buck, Erica Flapan

Proceedings of Symposia in Applied Mathematics
Volume **66**, 2009

A Brief Introduction to Knot Theory from the Physical Point of View

Colin Adams

ABSTRACT. In recent years, there has been an upsurge in interest in physical knot theory, which is the investigation of questions related to how a knot physically sits in Euclidean 3-space. This is motivated by applications that are concerned with exactly these questions. This article is an introduction to knot theory but with particular emphasis on those aspects of the theory that are relevant to physical knot theory. In particular, the last section focuses on superinvariants, which as of yet have not been fully explored.

1. Introduction

Knots have existed in some form or another for thousands of years. Initially, interest in knots came out of a desire to apply them in very practical situations. Early humans needed means to tie skins to the body for warmth. Sailors needed to have a variety of means to tie ropes to hoist sails. Surgeons needed to tie off sutures.

Eventually, in the late nineteenth century, the mathematical beauty of the field became apparent and pure mathematicians began to consider the mathematical theory of knots. There followed a period of mathematical exploration. But in recent years, we have seen a explosion of interest in the applications of knot theory.

In this article, we will discuss the basic background necessary to get started in knot theory. We will focus on those aspects relevant to the applications and to the subsequent articles in this volume. In particular, unlike most introductions to knot theory, we will be interested in those geometric quantities related to the physical realizations of knots in 3-space. In no sense is this an inclusive introduction, but rather, focuses on the author's particular interests. Much of the following material appears in greater detail in the books [**1**], [**8**], [**9**], [**10**], [**14**], [**20**],[**25**], [**26**], [**27**], and others, with the exception of the last section, which has not appeared before.

Take a piece of string. Tie a knot in it and then glue the two loose ends of the string together. The result is what we call a knot. We consider two knots to be equivalent if, when made out of string, one can be rearranged to look like the other without cutting the knotted loop of string open.

2000 *Mathematics Subject Classification.* 57M25.

Of course all of this can be made mathematically rigorous. A knot is an embedding of a circle in 3-space and two knots are equivalent if there is an ambient isotopy of 3-space that takes one knot to the other. However for our purposes, the intuition of a knotted loop of string in space will suffice.

We also consider a finite collection of knotted loops in space, which we call a link. Again, equivalence of links corresponds to a rearrangment of the one collection to another collection without passing any of the loops through one another or themselves.

Often in the literature, there is no explicit distinction made between a given embedding of a knot in space, called a **conformation**, and the equivalence class of conformations that are equivalent to it under ambient isotopy. However, for our purposes, it will be useful to make a distinction. For us, K will usually represent a fixed conformation of the knot and $[K]$ will represent the equivalence class of conformations for that knot, which is called the **knot type**. (However, as in common usage, we will often refer to the knot type as the knot.)

We represent a given knot in 3-space by a projection to a plane. In addition to the projection of the loop to the plane, we keep track of which strand goes over or under at each crossing. We avoid those projections that have singular points other than transverse double points, restricting ourselves to the so-called **regular projections**. See Figure 1.

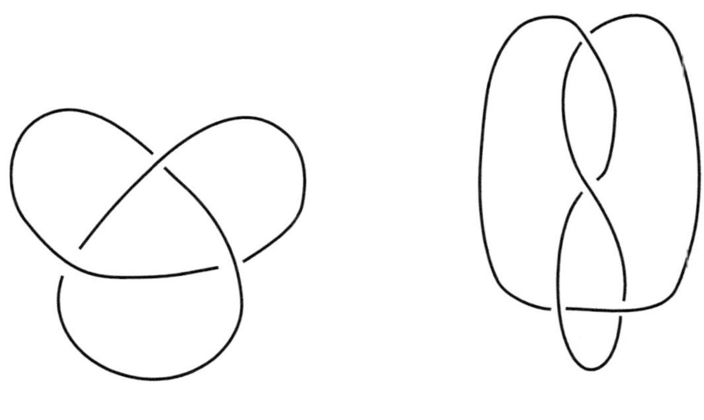

FIGURE 1. Projections of the trefoil and figure-eight knots.

Note that it is easy to obtain projections of a given knot with arbitrarily many crossings by simply twisting a part of the knot to create additional crossings. However, there must be a lower bound on the number of crossings for all the projections of a given knot.

DEFINITION 1. The **crossing index** of a knot K, denoted $c[K]$, is the least number of crossings for any projection of the conformations of that knot type.

In Figure 1, these projections are minimal crossing projections and therefore they yield the crossing indices for these particular knots. In general, however, it is difficult to determine the crossing index of a knot. To do so, one must find a projection with a given crossing number and then show that it cannot be the case that there are any projections of this knot with fewer crossings.

Crossing index is our first example of an invariant for knots. An **invariant** is a collection of mathematical objects, such as a collection of numbers, or polynomials, or groups, such that we associate one of the collection to each knot type. Two knots that are equivalent have the same object associated to them. In the case of crossing index, we are associating a nonnegative integer to each knot, and by definition, we have ensured that the result is an invariant, as we consider every possible projection of every possible conformation of the knot in the definition. Invariants of this type are typically easy to define but hard to calculate.

As we proceed, we will see additional examples of invariants defined in this manner. But we will also see invariants defined in other ways for which it is not immediately obvious that they are invariants. However, once we prove they are invariant, it is often substantially easier to determine the values of the invariant for a given knot type.

Define an orientation on a knot as a direction to travel around it, usually denoted by an arrow on the knot. Given two knots K_1 and K_2, we can define their **composition** $K_1 \# K_2$ as follows. Choose a projection of each and an arc on the outside of the projection that does not intersect the crossings. Then remove the arc from each knot and connect the ends in pairs as in Figure 2 by arcs that do not cross either projection or each other.

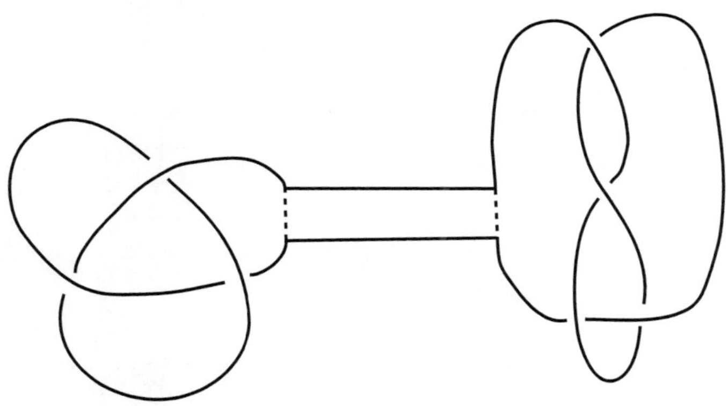

FIGURE 2. The composition of two knots.

Although it is not immediately obvious that this operation is well-defined, since we have a lot of choice of projections and arcs on those projections, it turns out that there are at most two distinct knots that can result, determined by whether or not fixed orientations on the two original knots match or do not match when the composition is constructed.

One might wonder whether or not the composition of two nontrivial knots must yield a nontrivial knot. In fact, the answer is yes. (See [**1**] for example, for a proof).

DEFINITION 2. A knot is **composite** if it is equivalent to the composition of two nontrivial knots. A nontrivial knot that is not composite is called **prime**.

Note that $c[K_1 \# K_2] \leq c[K_1] + c[K_2]$ since we can choose projections of K_1 and K_2 that realize their crossing indices and use them to create a projection of $K_1 \# K_2$ that has $c[K_1] + c[K_2]$ crossings. But we would like to know there is no simpler projection of the composition.

CONJECTURE 1. $c[K_1 \# K_2] = c[K_1] + c[K_2]$.

This simply stated conjecture has been open for over 125 years. It is known to be true for a few categories of knots, which we discuss shortly. But it would be extremely helpful to know this for all knots. In [**24**], it was announced that for any nontrivial knots K_1 and K_2, the following is true.

THEOREM 2. $\frac{c[K_1)]+c[K_2]}{152} \leq c[K_1 \# K_2] \leq [K_1] + c[K_2]$.

Although this does not appear so close to the full strength of the conjecture, it is a substantial first step.

One of the goals in defining invariants is to use them to distinguish between different knots. Over the last 120 years, mathematicians have been cataloguing the various knots and determining the number of prime knots of a given crossing number. Table 1 shows a list of what is known at this time. Note that this list of possible knots does not distinguish between a knot and its mirror image.

TABLE 1. Numbers of prime knots

Crossing Index	# of Prime Knots
0	1
1	0
2	0
3	1
4	1
5	2
6	3
7	7
8	21
9	49
10	165
11	552
12	2176
13	9988
14	46,972
15	253,293
16	1,388,705

Determining these numbers is extremely difficult. If we consider 16-crossing knots for instance, we must first consider all of the possible 16-crossing projections, where crossings are not yet specified. One can show that there are no more than 16! such diagrams. Then we multiply this by the number of choices for crossings, which is 2^{16}. Now, from amongst all of the projections that result, we must determine which can be simplified to knots with smaller crossing number, which are in fact composite knots and which are equivalent to which others or to their reflections. The number of 16-crossing knots was determined in 1997 in [**15**].

Our next invariant is a very natural one. It is another one that is easy to define but hard to calculate.

DEFINITION 3. The **unknotting index** of a knot, denoted $u[K]$, is the least number of crossing changes in any projection of the knot to obtain a projection of the trivial knot.

So, for instance, the unknotting number of the trivial knot is 0 and it is the only knot for which this is true. The unknotting number of the trefoil knot is 1. In fact, the unknotting number of the 7_4 knot is 2, however, proving this takes some work. One has to show that there cannot be any projection of the 7_4 knot that can be unknotted with one crossing change. (See $K_3 = 7_4$ in Figure 3.)

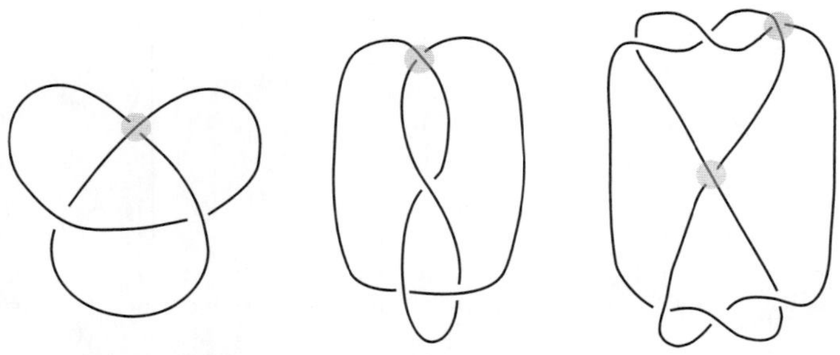

FIGURE 3. The unknotting number of knots.

How does unknotting index behave with respect to composition?

CONJECTURE 3. $u[K_1 \# K_2] = u[K_1] + u[K_2]$.

This conjecture has also been open for many years. In [**37**], it was proved that an unknotting index one knot is prime. In other words, $u[K_1 \# K_2] \geq 2$ for K_1 and K_2 nontrivial knots. This is a first step in proving this elusive conjecture.

Although results on unknotting index have been limited, recent work on Heegaard Floer homology by Peter Ozsváth and Zoltán Szabó has lead to some success, including the classification of all knots of nine or fewer crossings with unknotting index one and all alternating (to be defined shortly) knots of ten crossings with unknotting index one ([**33**]).

Our next invariant was introduced by H. Schubert in 1954. It has proved to be extremely useful. Fix a conformation of a knot K in 3-space. For each axis in space, there will be a projection of the entire knot conformation onto the axis, as in Figure 4. Relative to that projection, there will be some number of local maxima of the knot. We define the bridge number of the conformation to be the least number of local maxima that can be attained when projecting that conformation to any possible axis. We then minimize this over all conformations of the knot. Note that we need only consider axes through the origin and that we can represent each such axis by a unit vector from the origin ending at a point on the unit sphere S^2.

Thus, we can define the bridge index as follows:

DEFINITION 4. The **bridge index** of a knot is given by

$$b[K] = \min_{K \in [K]} \min_{\vec{v} \in S^2} (\# \text{ of local maxima}).$$

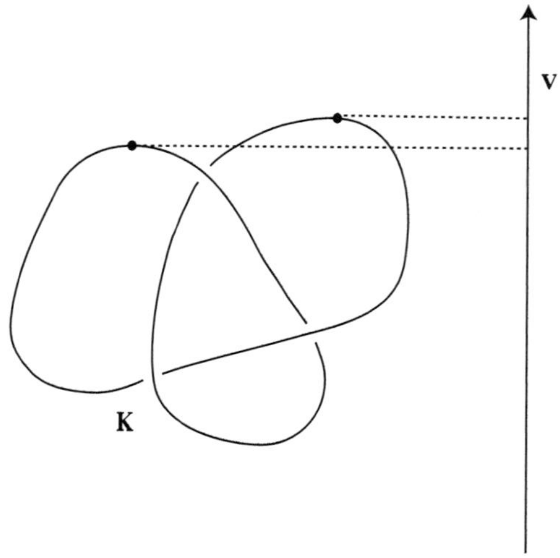

FIGURE 4. Counting local maxima when projecting to an axis.

For instance, the bridge index of the trivial knot is 1 and one can easily prove that this is the only knot with bridge index 1. The bridge indices of the trefoil and figure-eight knot are both 2.

We do know the behavior of bridge index with respect to composition (see [**38**] or [**39**]):

THEOREM 4. $b[K_1 \# K_2] = b[K_1] + b[K_2] - 1$.

The knots with bridge index two have turned out to be a relatively simple class of knots for which quite a bit is known. All such knots have projections similar to that appearing in Figure 5, with sequences of twists alternately in the center and on the right.

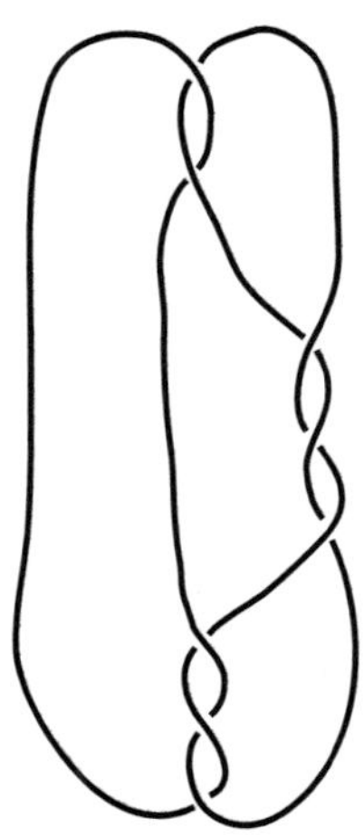

FIGURE 5. A two-bridge knot.

Two-bridge knots turn out to be a subclass of the alternating knots (see the next section).

Our next invariant has also proved to be quite useful. Given a projection of a knot, we say it is in braid form if there exists a point A in the plane off the projection such that as we travel along the knot in a given direction, the vector from A to the point on the knot always turns in the counterclockwise direction. (See Figure 6.)

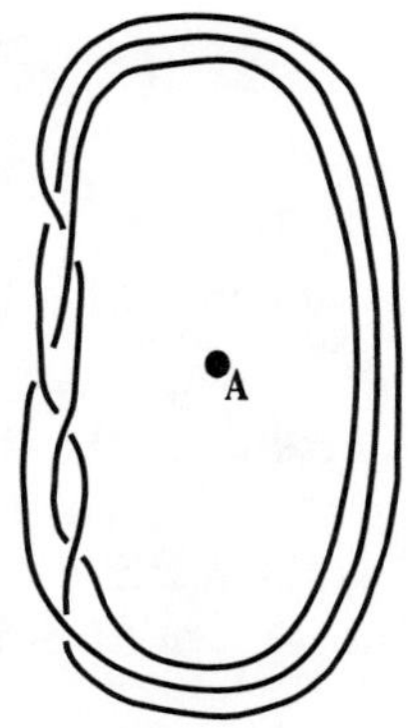

FIGURE 6. A projection in braid form.

One can show that every knot has braid projections.

DEFINITION 5. The **braid index** of a knot, denoted $\beta[K]$, is the least number of times that we turn counterclockwise around the point A in a braid projection of the knot.

Note that the only knot of braid index 1 is the trivial knot. Knots and links of braid index 2 must appear as in Figure 7, but with arbitrarily many crossings. It is only when the braid index gets up to three that the corresponding knots become interesting.

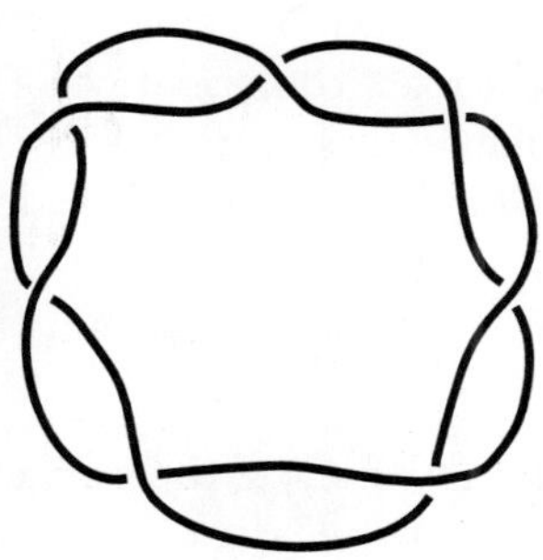

FIGURE 7. A 2-braid knot of 7 crossings.

It turns out that we can relate braid index to the so-called Seifert circles of the knot's projections. Given a projection of an oriented knot, we can hop on the knot and travel in the direction of the orientation. When we reach a crossing, we turn

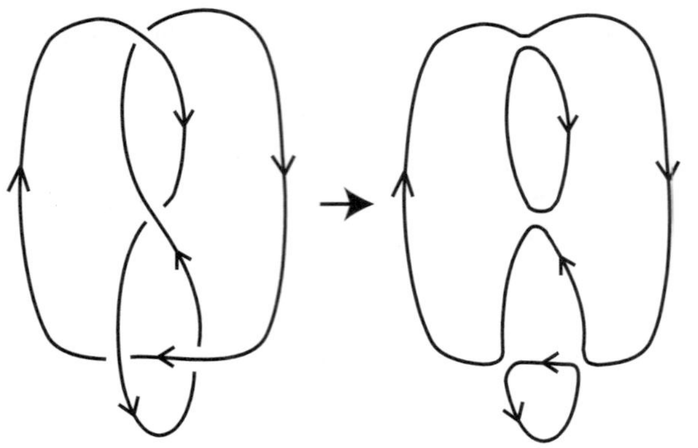

FIGURE 8. The Seifert circles obtained from an oriented projection.

off the knot onto the strand that crosses in the direction of the orientation. This has the effect of cutting the knot open at each crossing and regluing it without a crossing in a manner that preserves the orientation. The result will be a collection of simple closed curves in the plane, known as **Seifert circles**. See Figure 8.

THEOREM 5 ([**41**]). *The braid index of a knot is the least number of Seifert circles in any projection of the knot.*

This theorem allows an alternative definition of braid index which will prove useful in the last section of this paper.

It is straightforward to see that $\beta[K_1\#K_2] \leq \beta[K_1] + \beta[K_2] - 1$ by drawing a picture. It is not at all obvious that this is in fact an equality.

THEOREM 6 ([**6**]). $\beta[K_1\#K_2] = \beta[K_1] + \beta[K_2] - 1$.

A particular invariant that has a lot of intuitive appeal is the so-called rope length of a knot. The idea is to realize a particular knot by tying it in a rope and then gluing the ends of the rope together in a smooth manner. Depending on the radius of the rope, it will take a certain length of rope to achieve the knot. We can make this mathematically precise.

DEFINITION 6. Let K be a fixed conformation of $[K]$ in space. For each point $x \in K$, let $D(x, R)$ be the planar disk of radius R centered at x and orthogonal to K. Define $R = sup\{R : D(x, R) \cap D(y, R) = \varnothing \text{ for all } x \neq y \in K\}$ The **rope length of K**, denoted $L(K)$, is given by $L(K) = \frac{arclength(K)}{R}$. The **rope length of [K]** , denoted $L[K]$, is the minimum of this value over all conformations of the knot.

In general, this is a difficult invariant to determine. Experimental evidence suggests that the rope length of the trefoil is about 32.7 ([**35**]). In [**12**], quadrisecants are used to prove that a nontrivial knot must have rope length at least 31.3. (Note: Sometimes, rope length is defined by dividing arclength by diameter, in which case these numbers are halved.) But as of yet, the rope length of any nontrivial knot has yet to be determined.

The rope length of a conformation can be thought of as a so-called energy function on the knot. An energy function satisfies certain conditions one would expect from the energy associated wth a repulsive electrostatic charge spread along the knot.

(1) $E(K)$ varies continuously with deformations of the curve.
(2) $E(K)$ is invariant under Euclidean isometry and change of scale.
(3) $0 < E(K) < \infty$ for embedded K.
(4) As a knot approaches self-intersection, $E(K)$ approaches ∞.
(5) There are only finitely many knot types below any given energy and only trivial knots below some fixed energy level.

All of these conditions are satisfied by rope length, which is one of several possible energy functions that can be applied to knots.

There is another quantity related to the rope length of a knot conformation. Given a fixed knot in space, imagine surrounding it with a large sphere. Then for each point on the sphere, we can imagine labelling that point with the number of crossings we see from that vantage point. Ultimately, every point on the sphere corresponding to a regular projection is labelled with a positive integer. In fact, the sphere is subdivided into open subsets by a graph embedded on it, such that all points in each subset share the same label. The graph includes all the points where the viewpoint yields a projection that is not regular. We call the sphere with labels the **crossing sphere** for the conformation. We can then average the labels over the sphere to obtain what is known as the **average crossing number** of the conformation, denoted $acn(K)$.

Note that $c[K] \leq acn(K)$ for any conformation $K \in [K]$. Also note that we can obtain conformations of K that have average crossing number arbitrarily close to $c[K]$ by taking conformations that are very close to being flat in the projection that realizes the crossing index. These conformations have very few viewpoints that do not yield crossing number $c[K]$ and so their average crossing number is very close to $c[K]$.

Amazingly enough, rope length of conformations is related to average crossing number.

THEOREM 7 ([**7**]).
(1) $11L(K)^{\frac{4}{3}} \geq 4\pi acn(K)$,
(2) $\frac{1}{4}L(K)^2 \geq 4\pi acn(K)$.

Of course, this means that if we minimize over all projections, we obtain:

$$11L[K]^{\frac{4}{3}} \geq 4\pi c[K],$$
$$\frac{1}{4}L[K]^2 \geq 4\pi c[K].$$

A final invariant that we will consider in this section is the **stick index** of a knot, denoted $s[K]$, and sometimes also called the polygonal index. This is the least number of sticks glued end-to-end to form a conformation of the knot K in 3-space. It is of particular interest to biochemists who may be interested in constructing a knotted molecule out of atoms or other submolecules.

Clearly, the stick index of the trivial knot is 3, as a triangle will do the trick. The obvious first question we might ask is how many sticks it takes to make a nontrivial knot. Three sticks yields only a triangle, which is clearly trivial. Four sticks can yield at most one crossing, and therefore cannot generate a nontrivial knot. If one constructs a knot from five sticks, one could project down one of the sticks and see only four sticks in the plane, which yields at most one crossing, so again this cannot be a nontrivial knot. In Figure 9 , we see a trefoil knot constructed with six sticks. So if T is the trefoil knot, we know that $s[T] = 6$.

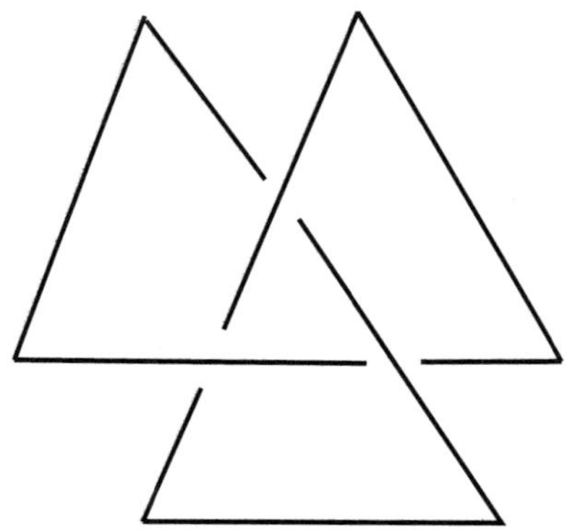

FIGURE 9. A trefoil knot realized with six sticks.

How does stick index behave under composition? We do know the answer in the case of composition of trefoil knots. Let nT represent the composition of n trefoil knots.

THEOREM 8 ([**2**]). $s[nT] = 2n + 4$.

This is somewhat surprising. Although the first trefoil costs us six sticks, each subsequent trefoil knot only requires an additional two sticks. The general case of how stick index behaves under composition is still open.

2. Types of knots

Certain categories of knots come up often and we know much more about them. In this section, we will discuss a few of those categories.

2.1. Torus Knots. A knot is said to be a **torus knot** if it is nontrivial and it can be realized on the surface of an unknotted torus in 3-space. Such a knot is called a (p, q)-torus knot, denoted $T_{p,q}$, if it wraps p times the short way around the torus (called the meridianal direction) and q times around the long way (called the longitudinal direction). See Figure 10. The trefoil knot is the $(3, 2)$-torus knot. In fact, a (p, q)-torus knot is also a (q, p)-torus knot.

Torus knots are a relatively simple category of knots, for which we know a reasonable amount of information. For instance, we know the following:

THEOREM 9. *Let $p < q$.*

(1) $c[T_{p,q}] = (p-1)q$ ([**32**]),
(2) $b[T_{p,q}] = p$ ([**38**]),
(3) $u[T_{p,q}] = (p-1)(q-1)/2$ ([**22**]),
(4) $s[T_{p,q}] = 2q$ *for* $2 \leq p < q < 2p$ ([**16**]),
(5) $s[T_{p,2p+1}] = 4p$ *for* $p \geq 2$ ([**5**]),
(6) $c[K_1 \# K_2] = c[K_1] + c[K_2]$ *when* K_1 *and* K_2 *are torus knots* ([**11**]).

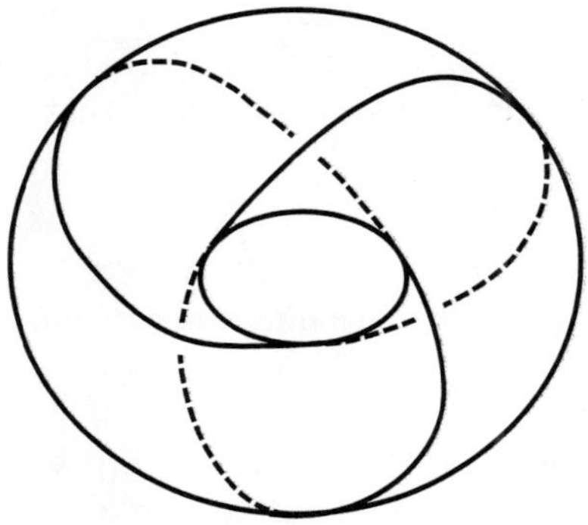

FIGURE 10. A trefoil knot realized as a torus knot.

Although we do know many of the stick indices of torus knots, we do not yet know the stick index for one of the simplest categories of torus knots, that being the 2-braids as in Figure 7.

2.2. Alternating knots.

DEFINITION 7. A knot or link K is **alternating** if it has a projection such that the crossings alternate from over to under as we travel along each of the components of K.

Many of the knots we have already considered are alternating. They include the trefoil knot, the figure-eight knot and all of the prime knots of seven or fewer crossings. However, low crossing number is deceiving in that as crossing number increases, the fraction of knots that are alternating shrinks toward 0. Already, once we reach 13 crossings, the number of non-alternating prime knots exceeds the number of alternating prime knots. On the other hand, as mentioned previously, all two-bridge knots are alternating.

One of the most important results about alternating knots and links was proved in 1988.

DEFINITION 8. A projection of a knot or link is **reduced** if there are no obvious ways to decrease crossing number as in Figure 11.

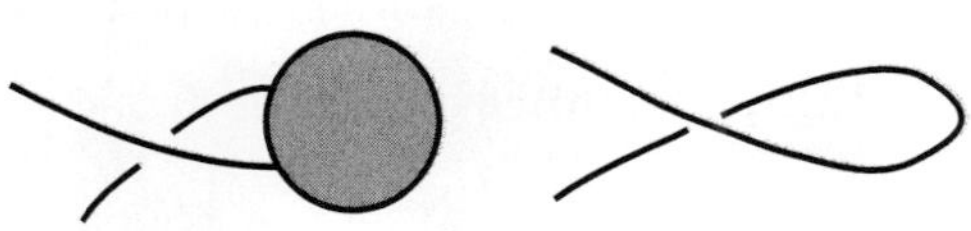

FIGURE 11. Projections that can be reduced.

THEOREM 10 ([**21, 31, 40**]). *The crossing index of an alternating knot or link is realized in any reduced alternating projection. A nonalternating projection of that knot or link must have a greater number of crossings.*

This is a very powerful result in that it allows us to easily read off the crossing index of alternating knots. It also has the following corollary.

FIGURE 12. A flype.

COROLLARY 11. *If K_1 and K_2 are alternating knots, then*

$$c[K_1 \# K_2] = c[K_1] + c[K_2].$$

This follows immediately from the easily checked fact that the composition of two alternating knots must itself be an alternating knot.

For over 100 years, it was conjectured that two reduced alternating projections of a given knot must be related through a sequence of flypes, as in Figure 12. In [**28**], this was proved to be true. Using this fact, the number of distinct prime alternating knots through 22 crossings was determined ([**34**]), showing in particular that for 22 crossings, there are 4,976,016,485 alternating knots. (Just in case you were wondering.)

3. Reidemeister moves and invariants

Given two projections of a knot, one would like to have a sequence of moves that gets us from the one projection to the other. In addition to the ability to deform a projection without changing the combinatorial pattern of regions in the plane, K. Reidemeister introduced three additional moves that change the diagram. These are the so-called Reidemeister moves. Each of the moves does not change the knot type, and only changes the projection locally. Each move can be performed in either direction. See Figure 13.

In 1932 ([**36**]), Reidemeister proved that two knot projections correspond to the same knot type if and only if they are related through ambient isotopy in the plane and a sequence of Reidemeister moves. This fact will allow us to prove that many invariants are in fact invariant.

As a first example, we consider the linking number of a 2-component oriented link. Orient each of the components. Choose a projection of the link. Label each crossing between the two different components with a $+1$ or -1 according to the convention that appears in Figure 14. Although for the purposes of linking number, we will only consider crossings between distinct components, the labeling convention will also apply to other crossings when we define writhe below.

DEFINITION 9. The **linking number** of the link L , denoted $lk[L]$, is given by

$$lk[L] = \frac{1}{2} \sum \pm 1$$

where the sum is over all crossings between the two distinct components.

Of course, this definition seems highly dependent on the projection we choose. However, it is straightforward to show this is not the case. We need only show that we would get the same answer if we used some other projection of the same link. In this case, since any two projections are related through Reidemeister moves, it is enough to show that none of the Reidemeister moves will change the result. First, note that Reidemeister move I does not impact the crossings between the two distinct components. It can only create or eliminate a crossing between a component

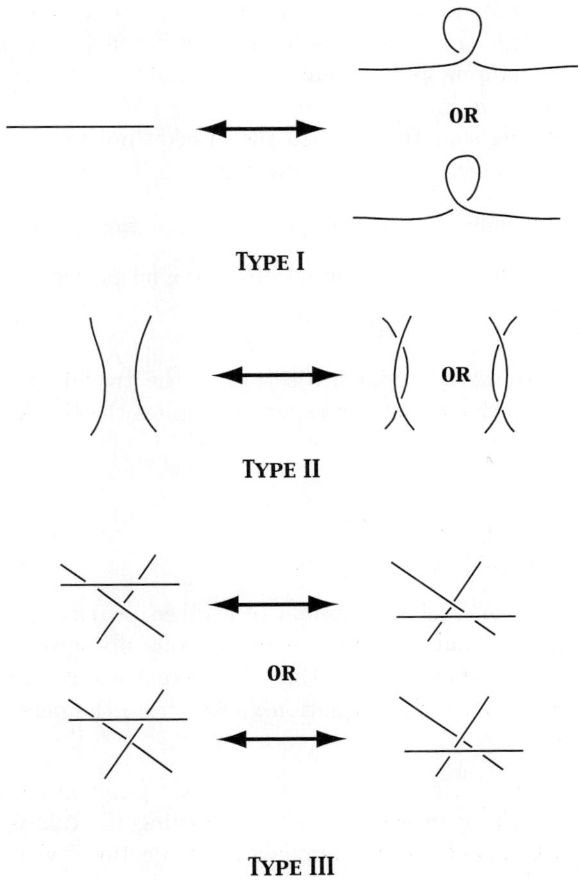

FIGURE 13. The Reidemeister moves.

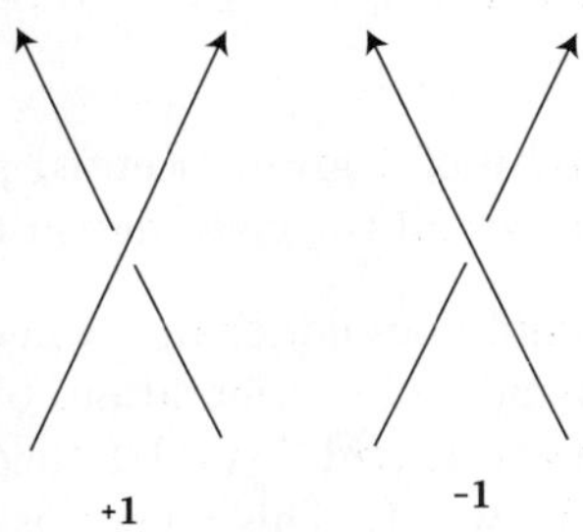

FIGURE 14. Labels for linking number.

and itself. Reidemeister move II can introduce or eliminate two crossings, however their labels always have opposite signs, so the sum of labels is unaffected.

Finally, the third Reidemeister move does not change the number of crossings or the labels on the crossings, and so the linking number is unaffected by such a move. Hence, we see that all projections of a given link must have the same linking number and the linking number is a well-defined invariant.

Now, let's try our hand at creating an invariant polynomial. This method is due to Lou Kauffman, and it results in the Kauffman polynomial $X(L)$ of an oriented link. It is a Laurent polynomial in the variable A, meaning that we have both positive and negative integer powers of A.

We start by defining something called the bracket polynomial of a projection. It is denoted by $<P>$ where P is the projection, and it satisfies three rules:

Rule 1: $<\bigcirc> = 1$ where $\bigcirc$ is a trivial knot projection with no crossings.

This says that the bracket polynomial of the trivial projection is 1.

Rule 2: $<P \cup \bigcirc> = (-A^2 - A^{-2})<P>$.

This says that adding a trivial projection of the trivial knot to an existing projection without introducing any crossings multiplies the bracket polynomial by a factor of
$-A^2 - A^{-2}$.

Rule 3: $<\times> = A<)(> + A^{-1}<\asymp>$,
$<\times> = A<\asymp> + A^{-1}<)(>$.

This says that the bracket polynomial of a given projection can be obtained from the bracket polynomials of the two projections obtained by eliminating a crossing in the two ways shown. Note that the second equation is not actually a separate fact. It is merely the first equation viewed from the perspective of having turned your head 90°.

These rules allow the computation of the bracket polynomial for a projection. We can inductively split open each crossing according to Rule 3 and obtain the polynomial for P in terms of the polynomials for projections with fewer crossings. Ultimately, once we have it down to projections with no crossings, we can use Rules 1 and 2 to find the polynomial we are after. Although one can show that it is invariant under Type II and Type III Reidemeister moves, the bracket polynomial does change under a Type I Reidemeister move. To remedy this problem we have one more definition.

DEFINITION 10. The **writhe** of a given oriented projection is the sum of the labels as defined in Figure 14 over all the crossings in the projection.

Note that the writhe of a knot does not change when the orientation is reversed. However, for a link, it does depend on the orientations of the individual components.

Writhe is unaffected by Type II and Type III Reidemeister moves. But Type I Reidemeister moves change it by ± 1. This allows us to use writhe to counteract the changes in the bracket polynomial caused by Type I Reidemeister moves.

We now define the Kauffman X polynomial as follows:

DEFINITION 11. The Kauffman polynomial of an oriented link L is obtained by fixing a projection P of L and then defining $X(L) = (-A^3)^{-w(P)} < P >$.

One can check that this polynomial is preserved under all three Reidemeister moves. Therefore, it does not depend on the particular projection, making it an invariant for knots and oriented links. In fact, if we replace A by $t^{-\frac{1}{4}}$, we obtain the original Jones polynomial, discovered by Vaughan Jones in 1985 in [**19**].

4. Superinvariants

In this section, we introduce variations on the traditional knot invariants that we have discussed so far. These are all physical knot invariants in the sense that they depend on how conformations of the knots sit in Euclidean 3-space. The initial invariant is due to Nicolaas Kuiper, who introduced superbridge index in [**23**].

DEFINITION 12. The **superbridge index** of a knot, denoted $sb[K]$ is given by

$$sb[K] = \min_{K \in [K]} \max_{\vec{v} \in S^2} (\# \text{ of local maxima}).$$

Notice that the only difference between the definitions of bridge index and superbridge index is that for a fixed conformation of the knot, we find the greatest number of local maxima coming from the projection to an axis, rather than the least number. We still then minimize this over all conformations.

From the definition, it is immediate that $b[K] \leq sb[K]$. In fact, Kuiper proved that $b[K] < sb[K]$ for any nontrivial knot. In [**4**], it was proved that $sb[K] \leq 2b[K]$.

Kuiper also showed that $sb[K] \leq 2\beta(K)$. Moreover, he determined the superbridge index for all of the torus knots.

THEOREM 12 ([**23**]). *If* $2 \leq p < q$, *then* $sb[T_{p,q}] = min\{2p, q\}$.

Questions about how superbridge index behaves under composition remain open. It is known that $sb[K_1 \# K_2] \leq sb[K_1] + sb[K_2]$([**4**]), but it is also known that the difference $sb[K_1] + sb[K_2] - sb[K_1 \# K_2]$ can be arbitrarily large([**17**]).

Jin has conjectured the following for nontrivial knots K_1 and K_2:

CONJECTURE 13. $sb[K_1 \# K_2] \leq sb[K_1] + sb[K_2] - 2$.

CONJECTURE 14. $sb[K_1 \# K_2] > max\{sb[K_1], sb[K_2]\}$.

There are a variety of other interesting questions that remain open about superbridge index. For instance, since $b[K] < sb[K]$, a nontrivial knot must have superbridge index at least 3. In [**18**], Jeon and Jin showed that there are only finitely many knots with superbridge index 3. The trefoil and figure-eight knots are known to have superbridge index 3. The only other candidates are $5_2, 6_1, 6_2, 6_3, 7_2, 7_3, 7_4, 8_4, 8_7$, and 8_9. (8_7 was eliminated in [**4**].) One would like to show that the other remaining knots actually have larger superbridge index and only the trefoil and figure-eight are superbridge index 3.

One of the primary reasons that we care about superbridge index is its relationship to stick index. Jin noted that if a knot is in a stick conformation, then when the conformation is projected to an axis, the local maxima must occur at vertices of sticks or along entire sticks perpendicular to the axis. In particular, this means that the number of local maxima obtained from a projection of this conformation to an axis is at most half the number of vertices. Since the number of vertices is equal to the number of sticks, this implies:

THEOREM 15. $s[K] \geq 2sb[K]$.

In the case of torus knots $T_{p,q}$ where $2 \leq p < q < 2p$, Thereom 12 then implies that $s[K] \geq 2sb[K] = 2q$. One can realize such a torus knot with $2q$ sticks as follows. Take two hyperboloids centered on the z-axis of differing waist size such that they intersect in a circle centered on the z-axis in the plane $z = 1$ and a circle of the same radius centered on the z-axis in the plane $z = -1$. The annuli that lie on the hyperboloids between the two planes form a torus with crescent moon crossection. Since the hyperboloid is a ruled surface, we can take q line segments on the outer hyperboloid that run from the bottom circle to the top circle. Then we take q line segments on the inner hyperboloid that connect the endpoints of the original segments on the top circle to the endpoints on the bottom circle. Pairs of segments from the two hyperboloids form meridians for the resulting knot. By adjusting the waists of the two hyperboloids, the appropriate number of longitudes can be attained. This yields Theorem 9(4).

Superbridge index is related to another invariant that dates back to [**13**] and [**29**].

DEFINITION 13. Let K be a polygonal knot. As we travel along the knot, we define the **exterior angles** to be those angles we turn through as we pass through each vertex. The **total curvature** of the conformation of the knot, denoted $tc(K)$, is the sum of these exterior angles. The total curvature of a smooth embedding of a knot is the least upper bound of the total curvatures of the polygons inscribed in it. The **total curvature of the knot type**, denoted $tc[K]$, is the greatest lower bound of the total curvatures of its conformations.

Milnor proved that for a nontrivial knot type $[K], tc[K] = 2\pi b[K]$, and for every fixed conformation, the total curvature is strictly greater than this lower bound. So, for instance, conformations of the trefoil knot can be constructed with total curvatures arbitrarily close to but slightly greater than 4π.

To see this, we mimic the crossing sphere but count local extrema rather than numbers of crossings. Assume K is a polygonal conformation of our knot. Taking a large sphere surrounding the knot, we label each point on the sphere with the number of local extrema we obtain when we project the knot to the axis corresponding to that point. Note that antipodal points receive the same label. The largest label must be at least as large as twice the superbridge index of the knot. The smallest label must be at least as large as twice the bridge index of the knot.

Since local extrema generically occur at vertices of the knot, we would like to determine which points on the sphere "see" a given vertex as a local extremum. Let v be a vertex of the knot, e_1 and e_2 the two edges of the knot meeting at v and α the exterior angle at v. The two planes through the origin that are perpendicular to e_1 and e_2 meet one another at angle α and cut the sphere into four regions. The two opposite regions, called loons, corresponding to the angle α are the two regions on the sphere where projections to the corresponding axis will generate a local extremum for v. The fraction of the sphere covered by these two loons is $\frac{2\alpha}{2\pi}$. Thus, the average number of local extrema, over all the choices of axes corresponding to points on the sphere is $\sum_{i=1}^{n} \frac{2\alpha_i}{2\pi}$. Therefore the average number of local maxima is half this, which is $\frac{tc(K)}{2\pi}$.

Thus, we see that $b(K) \leq \frac{tc(K)}{2\pi} \leq sb(K)$. In fact, since conformations do not exist with only one label on the entire sphere, these are actually strict inequalities. This implies that $tc[K] \geq 2\pi b[K]$ and $tc[K] \leq 2\pi sb[K]$. By applying linear transformations to squeeze a conformation to an axis, we can then show that $tc[K] = 2\pi b[K]$.

In the same manner that superbridge index can be obtained from bridge index, a variety of other invariants have "super" cousins. For instance, one can define crossing index as follows:

Fix a conformation of the knot in space. Consider all possible projections of that conformation to planes passing through the origin (the projection defined by a unit vector perpendicular to the plane). Take the least number of crossing obtained in this manner. Now minimize this over all conformations of the knot. The result is the crossing index.

DEFINITION 14. $c[K] = \min_{K \in [K]} \min_{\vec{v} \in S^2} (\# \text{ of crossings})$.

Of course, once we have defined it this manner, it becomes clear how we obtain the "super" version.

DEFINITION 15. The **supercrossing index** of K, denoted $sc[K]$, is given by

$$sc[K] = \min_{K \in [K]} \max_{\vec{v} \in S^2} (\# \text{ of crossings}).$$

We just maximize the number of crossings we see generated by a given conformation (known as the supercrossing number of that conformation), and then minimize this over all conformations. This is a natural invariant. For a given knot, what is the worst possibility for number of crossings from some point of view, if we can choose to take the best possible conformation?

Since the trivial knot can be placed in a plane, we see that its supercrossing index is 0. However for a nontrivial knot, the supercrossing index must be somewhat greater than the crossing index.

THEOREM 16 ([**3**]). *If K is a nontrivial knot, $sc[K] \geq c[K] + 3$.*

The idea of the proof is as follows. Travel along the knot in a particular direction. As we travel along, we extend out a tangent ray from the point we are at in the direction of travel. Since the knot is nontrivial, there must be a point along the knot when the ray re-intersects the knot somewhere. If we project in the direction of that ray, we obtain a non-regular projection. Slight changes in the projection direction will result in Type I and Type II Reidemeister moves on the corresponding projections. In other words, there will be projections of this conformation near one another that differ by at least three crossings. So every conformation must have a projection that has at least three more crossings than the crossing index for the knot.

However, determining supercrossing index seems to be difficult. For instance, since supercrossing index is at least 3 greater than crossing index, the supercrossing index of any nontrivial knot must be at least 6. In the case of the trefoil, there are conformations such that all of their projections have at most 7 crossings. So for the trefoil knot T, $6 \leq sc[T] \leq 7$.

However, it remains open at this time as to which of 6 or 7 is the supercrossing index of the trefoil knot.

We can relate supercrossing index to stick index.

LEMMA 16 ([**3**]). *Let $[K]$ be a nontrivial knot type.*

(1) $sc[K] \leq \frac{s[K](s[K]-3)}{2}$ *if $s[K]$ is odd.*
(2) $sc[K] \leq \frac{s[K](s[K]-4)}{2} + 1$ *if $s[K]$ is even.*

Note that if K is a fixed conformation of a knot, its supercrossing number will be the largest label that occurs on its crossing sphere. Its average crossing number will be the average of those labels. And the crossing index of the knot must be at most the smallest label that occurs on the crossing sphere. Note that the graph that appears on the crossing sphere consists of points on the sphere corresponding to non-regular projections. These non-regular projections correspond to transitional projections where Reidemeister moves take place. So each edge of the graph either corresponds to a Type I, a Type II or a Type III move. For the crossing sphere, the Type III edges are irrelevant since a Type III move does not change the number of crossings in the projections.

One can also create superinvariants out of other invariants obtained as minima over projections. For instance, one can define the superunknotting index as follows:

DEFINITION 17. The **superunknotting index** of K, denoted $su[K]$, is given by

$$su(K) = \min_{K\in[K]} \max_{\vec{v}\in S^2}(\#\text{ of crossings to trivialize } P).$$

Very little is known about this invariant. Note that there is also an "unknotting sphere" corresponding to this invariant. Each region in the complement of the graph is labelled with the number of crossings that must be changed to unknot the corresponding projections. Here, the Type I edges of the graph become irrelevant.

Similarly, one can define the superbraid index. Braid index can equivalently be defined as follows.

DEFINITION 18. The braid index of a knot or link is given by

$$\beta[K] = \min_{K\in[K]} \min_{\vec{v}\in S^2}(\#\text{ of Seifert circles in } P).$$

Hence superbraid index is defined as follows.

DEFINITION 19. The superbraid index of a knot or link K is given by

$$s\beta[K] = \min_{K\in[K]} \max_{\vec{v}\in S^2}(\#\text{ of Seifert circles in } P).$$

Once again, for a given conformation, we can define a "Seifert circle" sphere, the largest label of which will be at least as large as the superbraid index and the smallest label of which will be at least as large as the braid index. As of yet, there is little known about this construction.

These superinvariants offer many avenues for further research. There are numerous questions that remain to be answered.

References

[1] Adams, C. *The knot book. An elementary introduction to the mathematical theory of knots.* Revised reprint of the 1994 original. American Mathematical Society, Providence, RI, 2004.

[2] Adams, C., Brennan, B., Greilsheimer, D., and Woo, A., *Stick numbers and composition of knots and links*, J. Knot Theory Ramifications. 6, No. 2 (1997) 149-161.

[3] Adams, C., Othmer, J., Stier, A., Lefever, C., Pahk, S., Tripp, J., *An introduction to the supercrossing index of knots and the crossing map*, Knots 2000 Korea, Vol. 1 (Yongpyong), J. Knot Theory Ramifications 11 (2002), no. 3, 445-459.

[4] Adams, C., Agarwal, N., Allen, R., George, W., Winarski, R., Wootters, M., *Superbridge index of knots*, preprint, 2008.

[5] Adams, C., Shayler, T., *The projection stick index of knots*, to appear in the Journal of Knot Theory and its Ramifications, 2008.

[6] Birman, J., Menasco, W. *Studying links via closed braids. IV. Composite links and split links*, Invent. Math. 102 (1990), no. 1, 115–139. *Erratum*, Invent. Math. 160 (2005), no. 2, 447-452.

[7] Buck, G. and Simon, J., *Thickness and crossing number of knots*, Topology and its Applications, 91 (1999) 245-257.

[8] Burde, G., Zieschang, H. *Knots*, de Gruyter Studies in Math 5, Berlin-New York, 1985.

[9] Calvo, J., Millett, K., Rawdon, E. *Physical Knots: Knotting, Linking, and Folding Geometric Objects in* R^3, Contemporary Mathematics 304, American Mathematical Society.

[10] Cromwell, Peter R. *Knots and links*, Cambridge University Press, Cambridge, 2004.

[11] Diao, Y., *The additivity of crossing numbers*, J. Knot Theory Ramifications 13 (2004), no. 7, 857-866.

[12] Denne, E,. Diao, Y., Sullivan, J., *Quadrisecants give new lower bounds for the ropelength of a knot*, Geometry and Topology, 10 (2006) 1-26.

[13] Fáry, I., *Sur la courbure totale d'une courbe gauche faisant un noeud*, Bull. Soc. Math. France 77 (1949), 128-138.

[14] Flapan, E., *When topology meets chemistry. A topological look at molecular chirality*, Outlooks, Cambridge University Press, Cambridge; Mathematical Association of America, Washington, DC, 2000.

[15] Hoste, Jim, Thistlethwaite, Morwen, Weeks, Jeff, *The first 1,701,936 knots*, Math. Intelligencer 20 (1998), no. 4, 33–48.

[16] Jin, G. T., *Polygonal indices and superbridge indices of torus knots and links*, J. Knot Theory Ramifications 6, No. 2 (1997) 281-289.

[17] Jin, G. T., *Superbridge index of composite knots*, J. Knot Theory Ramifications 9 (2000), no. 5, 669-682.

[18] Jeon, C. B., Jin, G. T., *There are only finitely many 3-superbridge knots*, Knots in Hellas '98, Vol. 2 (Delphi). J. Knot Theory Ramifications 10 (2001), no. 2, 331-343.

[19] Jones, Vaughan F. R., *A polynomial invariant for knots via von Neumann algebras*, Bull. Amer. Math. Soc. (N.S.) 12 (1985), no. 1, 10–111.

[20] Kauffman, L. *On knots*, Annals of Mathematics Studies, 115. Princeton University Press, Princeton, NJ, 1987.

[21] Kauffman, L., *State models and the Jones polynomial*, Topology 26 (1987), no. 3, 395–407.

[22] Kronheimer, P. B., Mrowka, T. S., *Gauge theory for embedded surfaces. I*, Topology 32 (1993), no. 4, 773-826.

[23] Kuiper, Nicolaas, *A new knot invariant*, Matematische Annalen. 278, 193-209.

[24] Lackenby, M., *The crossing number of composite knots*, Thurston 60th Birthday Conference, Princeton, University, Princeton, NJ, June 10, 2007.

[25] Lickorish, W. B. R., *An introduction to knot theory*, Graduate Texts in Mathematics, 175. Springer-Verlag, New York, 1997.

[26] Livingston, C., *Knot theory*, Carus Mathematical Monographs, 24, Mathematical Association of America, Washington, DC, 1993.

[27] Manturov, V., *Knot theory*, Chapman and Hall/CRC, Boca Raton, FL, 2004.

[28] Menasco, W., Thistlethwaite, M., *The classification of alternating links*, Ann. of Math. (2) 138 (1993), no. 1, 113-171.

[29] Milnor, J., *On the total curvature of knots*, Ann. of Math. 52 (1950), 248-257.

[30] Murasugi, K., *Knot theory and its applications*, Birkhäuser Boston, Inc., Boston, MA, 2008.

[31] Murasugi, K., *Jones polynomials and classical conjectures in knot theory*, Topology 26 (1987), no. 2, 187-194.

[32] Murasugi, K., *On the braid index of alternating links*, Trans. Amer. Math. Soc. 326 (1991), no. 1, 237–260.

[33] Ozsváth, P., Szabó, Z., *Knots with unknotting number one and Heegaard Floer homology*, Topology, Vol. 44, Issue 4, July 2005, 705–745.

[34] Rankin, S., Flint, O., Schermann, J., *Enumerating the prime alternating knots. II.* J. Knot Theory Ramifications 13 (2004), no. 1, 101-149.

[35] Rawdon, E., *Approximating smooth thickness*, J. of Knot Theory and its Ramifications, 9 (2000) no. 1, 113-145.

[36] Reidemeister, K. *Knoten Theorie*, Springer-Verlag, 1932. English translation: Knot Theory, BCS Associates, 1983.

[37] Scharlemann, M., *Unknotting number one knots are prime*, Invent. Math. 82 (1985), no. 1, 37-55.
[38] Schubert, H., *Über eine numerische Knoteninvariante*, Math. Z. 61 (1954), 245-288.
[39] Schultens, J., *Additivity of bridge numbers of knots*, Math. Proc. Cambridge Philos. Soc. 135 (2003), no. 3, 539-544.
[40] Thistlethwaite, M., *A spanning tree expansion of the Jones polynomial*, Topology 26 (1987), no. 3, 297-309.
[41] Yamada, S., *The minimal number of Seifert circles equals the braid index of a link*, Inventiones Math. 89 (1987) 347-356.

Department of Mathematics, Bronfman Science Center, Williams College, Williamstown, MA 01267
E-mail address: cadams@williams.edu

Proceedings of Symposia in Applied Mathematics
Volume **66**, 2009

Topological Chirality and Symmetries of Non-rigid Molecules

Erica Flapan

ABSTRACT. We discuss the topology of non-rigid molecules. In particular, we introduce techniques which can be used to show that a molecule is topologically distinct from its mirror image, and we present the concept of the topological symmetry group of non-rigid molecules.

1. Introduction

Most chemists take a geometric approach to understanding the 3-dimensional structure of molecules. While this approach works well for rigid molecules, it does not always tell the whole story for flexible or even partially flexible molecules. In this chapter, we present a topological approach to the study of complex molecular structures.

In Section 2, we begin with an introduction to the concept of chemical chirality and its importance in chemistry. While chirality can be defined according to the behavior of a molecule in the laboratory, characterizing this concept mathematically can be subtle. The standard approach relies on the concept of *geometric chirality*, which treats all molecules as rigid objects. We present this approach in Section 3 together with examples of molecules where a geometric approach does not correspond to the chemical behavior of the molecules. In Section 4, we contrast chemical chirality and geometric chirality with the concept of *topological chirality*, which treats all molecules as completely flexible. Then in Section 5, we present four techniques for proving that a molecule is topologically chiral, and examples are given of when each technique can be applied. Finally, in Section 6, we discuss geometric and topological approaches to the study of molecular symmetry groups. In analogy with our study of chirality, we see that for complex molecules, the geometric approach given by using the *point group* does not always reflect molecular behavior as well as the topological approach given by the *topological symmetry group*.

2. Chemical chirality

Molecules can be modeled as graphs in space, where vertices represent atoms or groups of atoms, and edges represent bonds. For example, the molecule L–alanine can be illustrated by the graph in Figure 1. The dark triangular segment in the

2000 *Mathematics Subject Classification.* 57M25, 92E10.

figure indicates an edge which comes out of the plane of the paper towards the reader, the dashed segment indicates an edge which goes back behind the plane of the paper, and the ordinary line segments indicate edges which lie in the plane of the paper. In 3–dimensional space, the vertices of this graph lie at the corners of a regular tetrahedron.

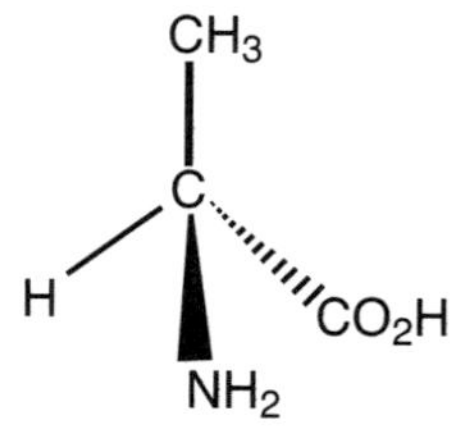

FIGURE 1. This molecule has the form of a tetrahedron.

The geometry of a rigid molecule determines many of its properties. However, large molecules may be flexible. For a flexible molecule, topology rather than geometry determines many of its properties. An example of a topologically interesting flexible molecule is the molecular Möbius ladder, which was *synthesized* (i.e. made in the laboratory) by David Walba [**24**] in 1982. This molecule (illustrated in Figure 2) has the form of a Möbius strip that is made of a 3-rung ladder. The rungs of the ladder are carbon-carbon double bonds and the sides of the ladder are chains of carbon atoms (indicated in the figure with a corner) and oxygen atoms.

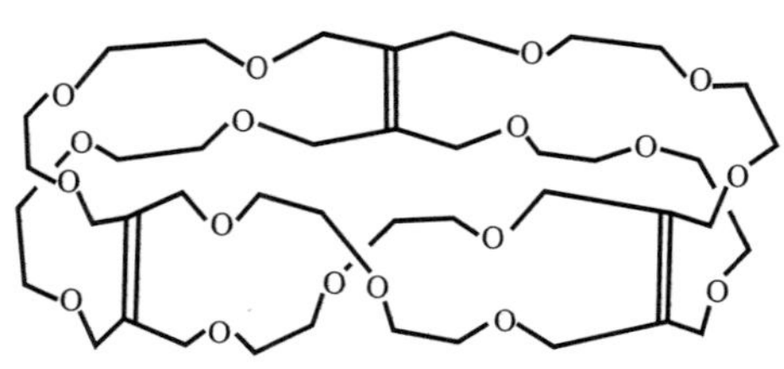

FIGURE 2. A molecular Möbius ladder.

While a molecular Möbius ladder has not resulted in new pharmaceutical products or other practical applications, synthesizing this molecule was an important chemical acheivement. Chemists create "targets of synthesis" such as a Möbius ladder, a knot, or a link in order to force the profession to develop new techniques of synthesis. Knotted and linked molecules were targets of synthesis since the early 1900's. Both have now been achieved. Dietrich-Buchecker and Sauvage [**7**] synthesized the first knotted molecule in 1989. This molecule is illustrated in Figure 3.

The molecular Möbius ladder was synthesized by forcing the ends of a ladder to join. Some of the ladders were joined without a half-twist and others were joined with a half-twist. Figure 4 illustrates a closed circular ladder without a half-twist.

These molecules are too small to see in a microscope, so it was not obvious that any of the closed ladders had the form of a Möbius ladder. In order to show that at least some contained a half-twist, Walba showed experimentally that some of the molecules were chemically different from their mirror image. Since a closed circular ladder with no twists is the same as its mirror image, not all of the molecules could

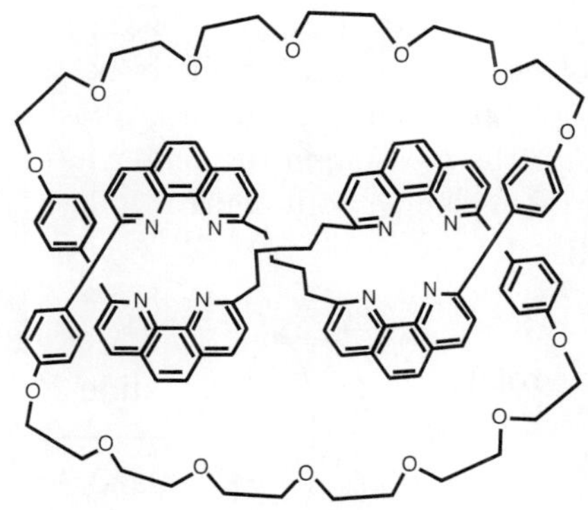

FIGURE 3. The first knotted molecule.

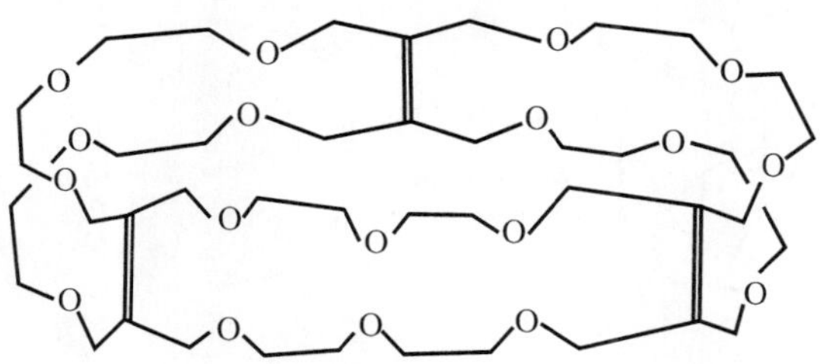

FIGURE 4. A closed ladder without any twists.

be such circular ladders. However, Walba did not know whether a molecular Möbius ladder was actually different from its mirror image. We illustrate a molecular Möbius ladder and its mirror image in Figure 5. A Möbius strip with a left-handed twist cannot be deformed to a Möbius strip with a right-handed twist. But a Möbius ladder has deformations that a Möbius strip does not have. So perhaps a Möbius ladder could be deformed to its mirror image, in which case perhaps a molecular Möbius ladder could convert itself to its mirror form.

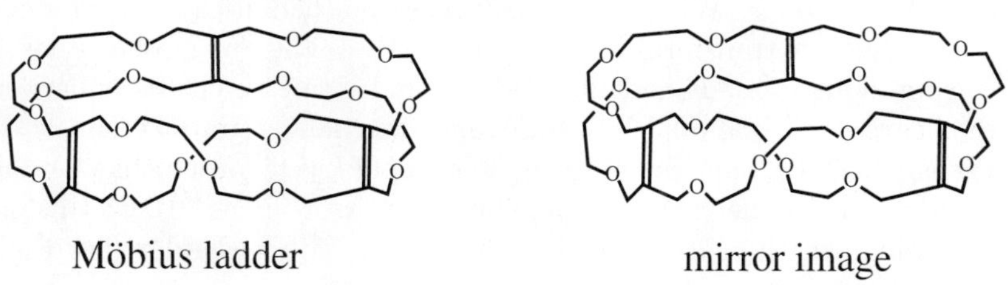

FIGURE 5. A molecular Möbius ladder and its mirror image.

In fact, it is possible to deform a Möbius ladder to its mirror image. We illustrate such a deformation in Figure 6. For simplicity we represent the Möbius ladder as a graph without indicating the different types of atoms. We number the atoms just to help us keep track of the deformation. The first step of the deformation is to rotate the graph by 90°. Next, we slide the edges $\overline{25}$ and $\overline{36}$ forward. Then we slide vertex 5 up and pull the edge $\overline{34}$ to the left. Then we shorten the edge $\overline{16}$. Next we make the loop $\overline{1456321}$ into a figure eight. Finally, we fold the top and bottom of the figure eight back, and shorten the edges $\overline{43}$, $\overline{25}$, and $\overline{16}$. If we ignore the numbers, the final illustration is identical to the original illustration except the crossing in the center has been reversed. Thus this deformation takes the Möbius ladder to its mirror image.

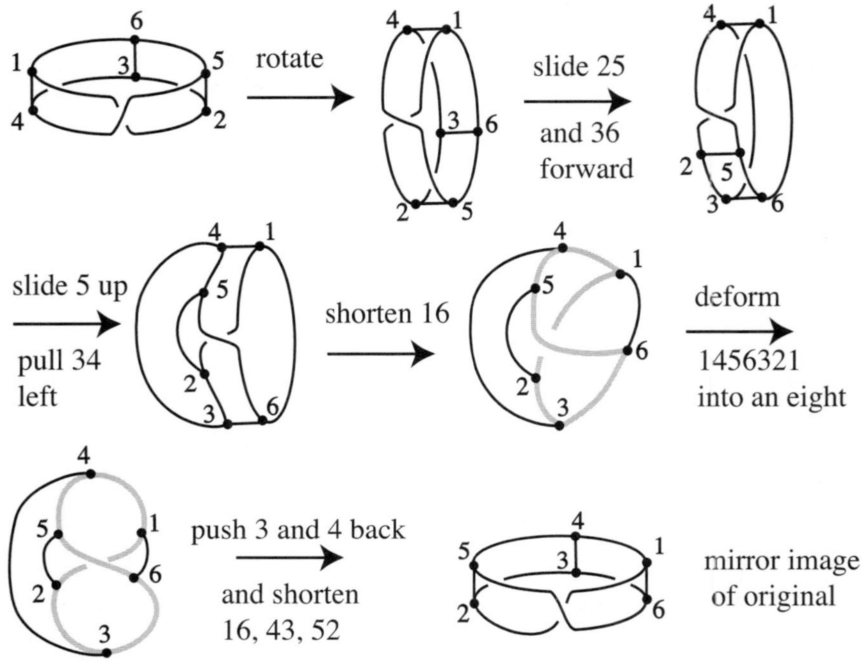

FIGURE 6. A deformation from a Möbius ladder to its mirror image.

Observe that the deformation illustrated in Figure 6 takes the rungs $\overline{14}$ and $\overline{36}$ to sides of the ladder. While this makes sense mathematically, from a chemical point of view such a transformation is impossible. The rungs of the ladder are carbon-carbon double bonds and the sides of the ladder are chains of carbons and oxygens. A carbon-carbon double bond cannot be changed into a chain of carbons and oxygens. In 1986, Jon Simon [**19**] used topology to prove that a molecular Möbius ladder cannot chemically change into its mirror image. To do this, he used colors to distinguish the sides from the rungs of a Möbius ladder (see Figure 7). Then he proved that this colored graph cannot be deformed to its mirror image.

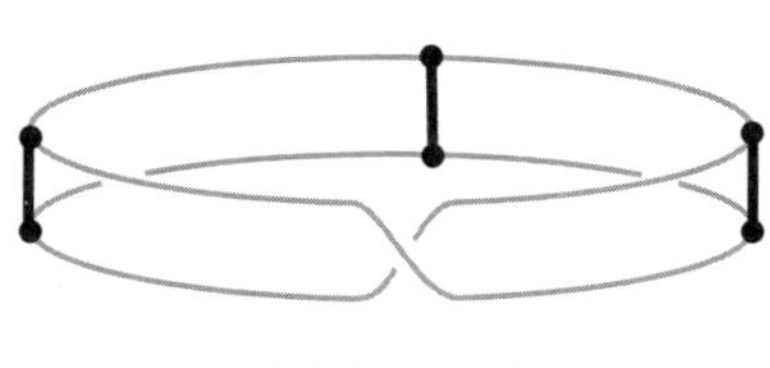

FIGURE 7. We represent a Möbius ladder and as a colored graph in order to distinguish sides from rungs.

Knowing whether or not a molecule is different from its mirror image is important in predicting its chemical behavior. Thus it is useful to introduce the following terminology.

DEFINITION 2.1. We say a molecule is **chemically achiral** if it can change into its mirror image. Otherwise, we say it is **chemically chiral**.

Chemists omit the word "chemically" and just say *chiral* and *achiral*. We include the word *chemically* to make it clear that this is an experimental rather than a mathematical definition. The word "chiral" comes from the ancient Greek word for hand because a hand cannot change into its mirror image. Thus the word "achiral" means not like a hand.

Living organisms generally have a preferred handedness on both a large and a small scale. For example, the human heart is normally on a person's left side, and the amino acids in the human body are chemically chiral. Your right hand reacts differently to a right glove and a left glove. Similarly, organisms react differently to the two forms of a chiral molecule. For example, one form of the molecule carvone smells like spearmint and the other form smells like caraway. Generally, one form of a medication is more effective and the mirror form has more side effects. Occasionally the two mirror forms of a pharmaceutical product have different uses. For example the medication Darvon is a painkiller, while its mirror form, Novrad, is a cough medicine.

If a pharmaceutical product is known to be chiral, then the pharmaceutical company that produces it may choose to manufacture only the preferred form in order to minimize side effects and maximize effectiveness.

3. Geometric chirality

The concept of chirality is so important that we will spend much of this article discussing how chirality can be characterized and recognized mathematically. Such a characterization is useful because it enables chemists to know whether or not a molecule will be chiral even before it is synthesized. However, there is no mathematical definition of chirality that perfectly corresponds to chemical chirality, since whether or not a molecule is chemically chiral often depends on how flexible the molecule is. In the next few sections we will introduce and compare two different approaches to this problem: a *geometric* approach which treats all molecules as rigid objects, and a *topological* approach which treats all molecules as completely flexible. In order to see the subtleties of characterizing chirality we will consider some molecules which are partially flexible.

We begin by discussing the geometric approach to chirality that was first introduced by Lord Kelvin in 1884. In particular, Lord Kelvin wrote:

> "I call any geometrical figure or group of points *chiral*, and say it has *chirality*, if its image in a plane mirror, ideally realized, cannot be brought to coincide with itself."

According to this definition, an achiral object is one which is identical to its mirror image. For example, a person is chiral because a person is not identical to its mirror image. The molecule and its mirror image, illustrated in Figure 8, each have the form of a tetrahedron. Since the groups of atoms at the four corners of the tetrahedron are all distinct, this molecule is different from its mirror image. Hence it is chiral according to Lord Kelvin's definition.

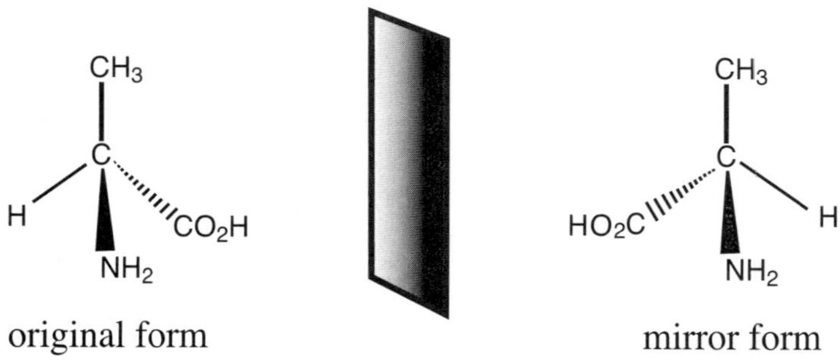

FIGURE 8. A chiral molecule and its mirror image.

Most modern organic chemistry textbooks reword Lord Kelvin's original definition only slightly. Here is the definition given in a standard organic chemistry textbook.

DEFINITION 3.1. An **achiral** molecule is one which is "superposable" with its mirror image. Otherwise, it is **chiral**.

We restate this mathematically as follows.

DEFINITION 3.2. A molecule is **geometrically achiral** if, as a rigid object, it can be superimposed on its mirror image. Otherwise it is **geometrically chiral**.

We add the word "geometrically" here to distinguish this definition from our earlier experimental definition, and to make it clear that this type of chirality comes from the geometry of a rigid molecule.

Many chemists consider geometric chirality and chemical chirality as synonyms. However there are examples of molecules where these two concepts do not coincide. In order to illustrate the distinction between these concepts, in 1954, Kurt Mislow [**17**] synthesized the molecule illustrated in Figure 9.

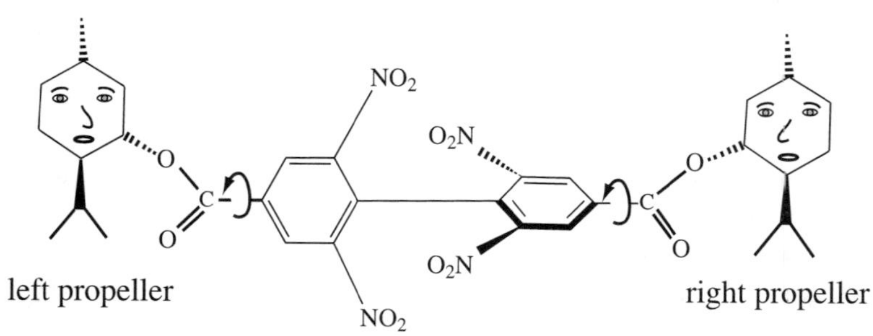

FIGURE 9. A molecule which is chemically achiral but geometrically chiral.

The two ends of the molecule, which we call "propellers", rotate simultaneously as indicated by the arrows. Of course the actual molecule does not have faces on it. The faces are drawn there to help illustrate the difference between the two propellers. The left propeller can be thought of as a person facing you with her left hand forward, as indicated by the dashed bond connecting the hexagon face to the oxygen. The right propeller can be thought of as a person facing you with her right hand forward. Because of the rigidity of the structure of the propellers, the notion of "forward" is unambiguous. The faces help us remember that forward is the direction the hands are pointing in. As rigid structures, a right propeller can never become a left propeller. If they were not rigid, we could flatten them out, and then they would be interchangeable. The hexagon next to the left propeller is vertical, while the hexagon next to the right propeller is horizontal.

We illustrate the mirror image of Figure 9 in Figure 10. Observe that the mirror form is identical to the original form except that the horizontal and vertical hexagons in the middle have been switched.

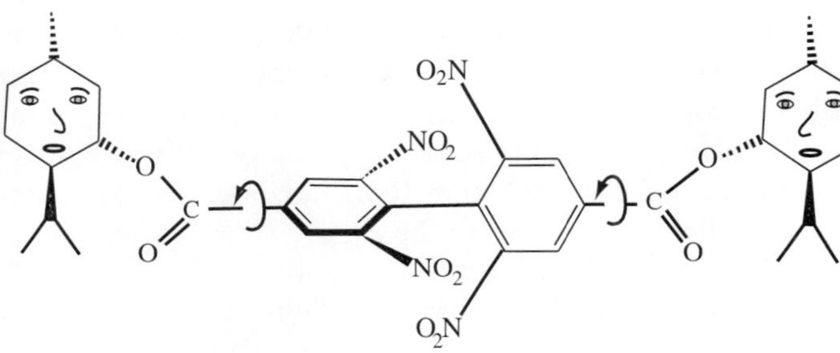

FIGURE 10. The mirror image of Figure 9.

We see as follows that the molecule illustrated in Figure 9 is chemically achiral. First we rotate the molecule about a horizontal axis which goes from left to right. This rotation takes the vertical hexagon in the middle to a horizontal hexagon, and takes the horizontal hexagon in the middle to a vertical hexagon. The propellers become horizontal as a result of this rotation. However, we can let the propellers rotate back to their original vertical positions. This gives us the mirror form illustrated in Figure 10. The motions we have just described can occur chemically, hence the molecule is chemically achiral.

On the other hand, we can see that the molecule in Figure 9 is geometrically chiral as follows. If we consider the molecule as a rigid object, then the propellers cannot rotate. In the original form, the left propeller is parallel to the adjacent hexagon. In the mirror form, the left propeller is perpendicular to the adjacent hexagon. As rigid objects, a left propeller cannot change into a right propeller. Hence as rigid objects, the original form and the mirror form are distinct. Thus the molecule is geometrically chiral.

Observe that as the propellers simultaneously rotate, they never get into identical positions relative to their adjacent hexagons. For example, if the left propeller is 45° behind its adjacent hexagon, then the right propeller is 45° in front of its adjacent hexagon. Thus, no matter what angle the propellers are at, the above argument shows that the molecule cannot be rigidly superimposed on its mirror image. Hence allowing the propellers to rotate will not enable the molecule to become geometrically achiral.

We see as follows that in the above sense this molecule and its mirror image are similar to a pair of rubber gloves. A right handed rubber glove is the mirror image of a left handed rubber glove. Also a left handed rubber glove can become a right handed rubber glove if it is turned inside out. However, at no stage of the turning inside out process is the rubber glove rigidly superimposable on its mirror image. So at no stage does it become geometrically achiral. These characteristics of a rubber glove motivated the chemist, Van Gulick [**23**] to introduce the following definition.

DEFINITION 3.3. A molecule is said to be a **Euclidean rubber glove** if it is chemically achiral, but it cannot attain a position which can be rigidly superimposed on its mirror image.

The word *Euclidean* indicates that geometric (i.e., physical or chemical) constraints are keeping the molecule from attaining a position which could be rigidly superimposed on its mirror image. The molecule illustrated in Figure 9 and a real rubber glove are both examples of Euclidean rubber gloves. We have seen that each can be transformed into its mirror image. If the molecule in Figure 9 were completely flexible, it could lie in the plane and become rigidly superimposable on its mirror image. However, the rigidity of the chemical bonds prevents it from lying in a plane. Similarly, if a rubber glove were 100% flexible, it could be stretched out flat like a pancake, and then it too would be rigidly superimposable on its mirror image. However, the physical limitations of a rubber glove do not allow enough stretchiness to completely flatten it out (try it!). Thus in both cases it is the physical limitations of the object which prevent it from attaining a position which is rigidly superimposable on its mirror image.

Walba introduced the following definition in order to see what happens if we ignore all of these types of physical limitations.

DEFINITION 3.4. A molecule is said to be a **topological rubber glove** if it is chemically achiral, but it cannot be deformed to a position which can be rigidly superimposed on its mirror image (even if it is completely flexible).

The molecule illustrated in Figure 9 is not a topological rubber glove, because if it were completely flexible it could be deformed to a position which could be rigidly superimposed on its mirror image. At first glance it might seem like a topological rubber glove would be impossible. If a molecule can be deformed to its mirror image should not it be possible to deform it to a position which can be rigidly superimposed on its mirror image? In fact, a topological rubber glove is possible. The first topological rubber glove was synthesized by Chambron, Sauvage, and Mislow [**5**] in 1997. This molecule is a pair of linked rings, where the pair of adjoined hexagons at the top can rotate. We illustrate the molecule and its mirror image in Figure 11.

We see as follows that this molecule is chemically achiral. The bottom ring of the original molecule can turn over, in which case it looks like the bottom ring in the mirror form. When the ring is turned over we write the H_3C as CH_3 to indicate that the tail is still attached at the C. When the pair of adjoined hexagons at the top of the original form rotates, it gets into the position of the two hexagons illustrated in the mirror form. In this way, we see that the original form can change into the mirror form. Since these motions are chemically possible, the molecule is chemically achiral.

On the other hand, it can be shown that even if the molecule in Figure 11 were completely flexible it could not be deformed to a position which could be rigidly

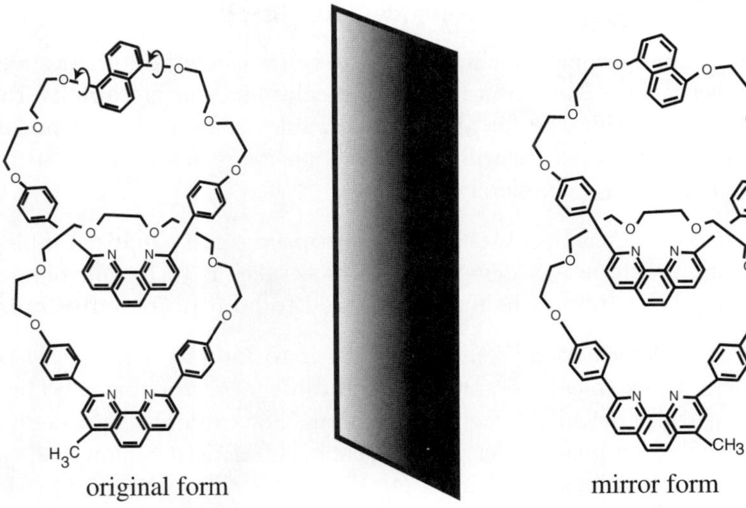

FIGURE 11. A topological rubber glove and its mirror image.

superimposed on its mirror image. We give a vague idea of the proof here (for a detailed proof see [**8**]). The H_3C in the original form gives the bottom ring an orientation. If the molecule were rigid, then the hexagons at the top cannot rotate, and hence they give the top ring an orientation. Thus we can model this molecule as a pair of oriented linked circles (as shown in Figure 12) with arrows to indicate an orientation on each circle. It can be shown that there is no deformation from this oriented link to its mirror image. It follows that there is no deformation from the molecule in Figure 11 to a position which can be rigidly superimposed on its mirror image. Hence the molecule in Figure 11 is a topological rubber glove.

FIGURE 12. This oriented link cannot be deformed to it mirror form.

We summarize the relationship between geometric and chemical chirality as follows. If a structure can be rigidly superimposed on its mirror image, then it is physically the same as its mirror image. Hence it must be chemically the same as its mirror image. Thus geometric achirality implies chemical achirality. Taking the contrapositive of this statement, we see that chemical chirality implies geometric chirality. On the other hand, Euclidean and topological rubber gloves are examples of molecules which are geometrically chiral but not chemically chiral. Hence geometric chirality does not imply chemical chirality.

4. Topological chirality

When we defined geometric chirality, we were treating all molecules as if they were completely rigid. There is no mathematical definition of chirality that corresponds to chemical chirality for all molecules, since some molecules are more or less rigid than others. Now we will take the opposite point of view and treat all molecules as if they were completely flexible.

DEFINITION 4.1. A molecule is said to be **topologically achiral** if, assuming complete flexibility, there is a deformation which takes it to its mirror image. If there is no such deformation, the molecule is said to be **topologically chiral**.

If there is no deformation from a molecule it to its mirror image, then there is no way for it to chemically transform itself into its mirror image. Thus every topologically chiral molecule is chemically chiral. For example, in order to prove that the molecular Möbius ladder is chemically chiral, Jon Simon [**19**] actually proved that it was topologically chiral.

The molecule illustrated in Figure 13 is chemically chiral because it has the form of a rigid tetrahedron. On the other hand, if the molecule were completely flexible we could twist around the edge with the H on the end and the edge with the HO_2C on the end. This would transform the original form of the molecule to the mirror form. Thus this molecule is topologically achiral. However, such a transformation is chemically impossible. This example, illustrates that topological achirality does not imply chemical achirality.

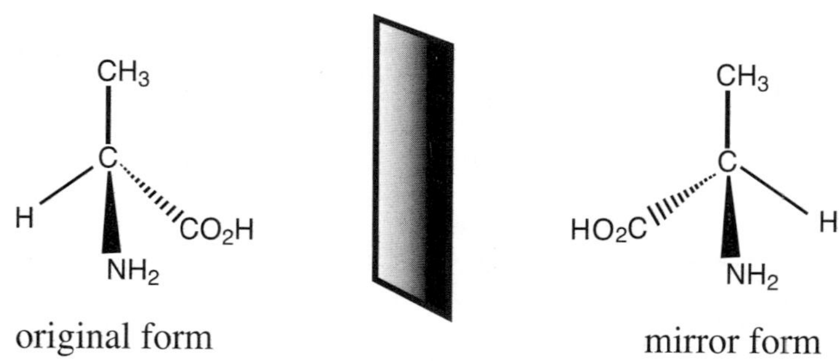

FIGURE 13. This molecule is chemically chiral because it has the form of a rigid tetrahedron.

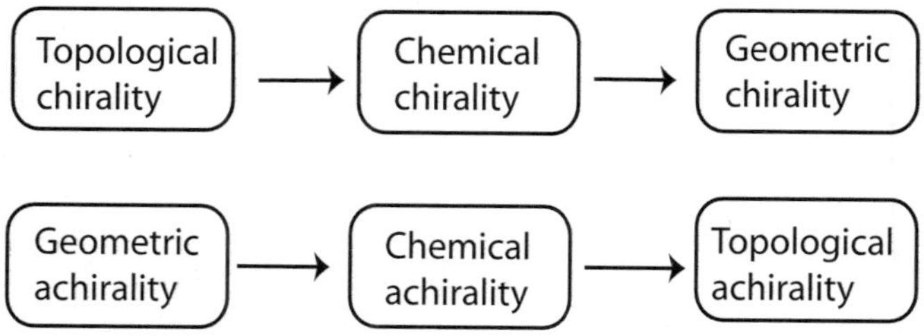

FIGURE 14. The relationship between the different types of chirality and achirality.

Figure 14 summarizes the relationships between the different types of chirality and achirality that we have been discussing. As we have seen in this and the previous section, none of the reverse implications hold.

If we heat a geometrically chiral molecule enough, it will change into its mirror form. By contrast, in order for a topologically chiral molecule to change into its mirror form, bonds would have to pass through one another. Even if we heat such a molecule, this will not occur. In this sense, topological chirality is a deeper and more enduring property than geometric chirality. This is one of the reasons that topological chirality is a meaningful concept for chemists.

5. Techniques to prove topological chirality

Now that we understand topological chirality and why it is chemically useful, we will introduce four different mathematical techniques for proving that a molecule is topologically chiral. Different techniques can be used for different molecules. There is no single technique that works for all molecules.

5.1. Knot polynomials. The first technique that we introduce involves using knot polynomials to show that a knotted or linked molecule is topologically chiral. As mentioned at the beginning of this paper, knotted molecules have been synthesized since 1989 (see Figure 3). The idea is that there is a method of assigning a polynomial to a knot and link such that if two knots or links have different polynomials, then one cannot be deformed into the other. In particular, if the polynomial assigned to a knotted molecule is different from the one assigned to its mirror image, then we know that the molecule is topologically chiral.

We present a very brief history of knot polynomials before we give an example. J. W. Alexander [**1**] developed the first knot polynomial in 1928. Alexander's polynomial played a major role in enabling topologists to distinguish one knot from another. In 1967, John H. Conway [**6**] discovered a recursive method for computing a related polynomial, in terms of the polynomials of simpler knots and links. While the Alexander polynomial and Conway polynomial are very important topological tools, they cannot help us detect topological chirality because the polynomials of a knot and its mirror image are the same.

In 1985, Vaughan Jones [**15**] developed a new polynomial which uses a recursive method similar to Conway's for its computation. In contrast with the Alexander and Conway polynomials, the Jones polynomial of a knot or link is often different from that of its mirror image. For example, Figure 15 illustrates a molecular knot and its mirror image together with their Jones polynomials. Observe that the picture of the mirror form of the molecule is identical to that of the original form except that the three crossings between bonds have been reversed. The Jones polynomial of the original form is $t + t^3 - t^4$ and the Jones polynomial of the mirror form is $t^{-1} + t^{-3} - t^{-4}$. Since these polynomials are different, we conclude that there is no deformation taking the original form its mirror form. Thus this molecular knot is topologically chiral. You may have noticed that the Jones polynomial is not technically a polynomial, since the exponents are negative as well as positive integers. However, it is standard to abusive terminology and refer to this as a polynomial.

Observe that for the knotted molecules illustrated in Figure 15, the Jones polynomial of the mirror form is identical to the Jones polynomial of the original form

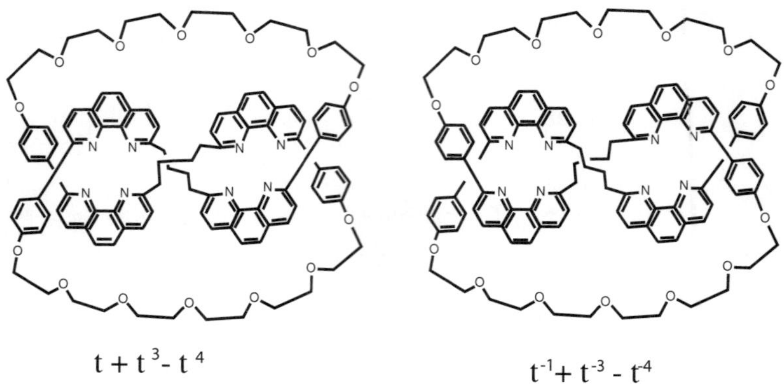

FIGURE 15. The Jones polynomial of a molecular knot and its mirror image.

except that the signs of the exponents have been reversed. This is not a coincidence. If K is any knot and K^* is its mirror image, then the Jones polynomial of the knot K^* will always be equal to the Jones polynomial of the knot K except that the signs of the exponents will be changed from positive to negative and from negative to positive. So if the Jones polynomial of a knot is not symmetric with respect to the signs of the exponents, then we can conclude that the knot is topologically chiral. Note however, that if the Jones polynomial of a knot is symmetric with respect to the signs of the exponents, it does not necessarily mean that the knot is topologically achiral. In particular, the Jones polynomial is useful for proving topological chirality, but not for proving topological achirality.

Using the Jones polynomial is a very effective way to prove that certain molecular knots and links are topologically chiral. However, this method will not help us establish chirality if a molecule does not contain any knots or links. For example, the molecular Möbius ladder (illustrated in Figure 2) does not contain any knots or links. Thus the Jones polynomial will not help us prove that it is topologically chiral. We will use the next method to prove that the molecular Möbius ladder is topologically chiral.

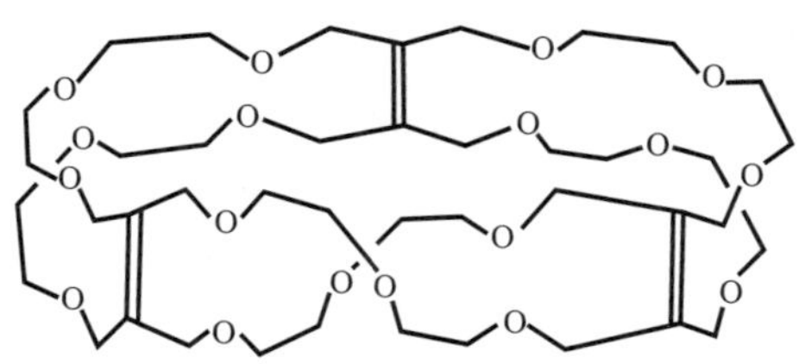

FIGURE 16. We cannot use polynomials to prove that the molecular Möbius ladder is topologically chiral.

5.2. 2-fold branched covers. The next method we introduce, involves using 2-fold covers to translate the question of whether a graph in 3-space is topologically chiral into the question of whether a particular link is topologically chiral. Covering spaces and branched covers have been a useful tool in topology for over 100 years.

However, Jon Simon [**19**] was the first to use this idea to prove that a molecule was topologically chiral.

In order to avoid getting too technical, we will not formally define a 2-fold branched cover (chapter 3 of [**8**] provides the definition of a 2-fold branched and gives a detailed exposition of Simon's proof). Instead, we present an intuitive picture of what a 2-fold branched cover is and how Simon used this idea to prove that the molecular Möbius ladder is topologically chiral.

We begin by representing the molecular Möbius ladder as a colored graph with different colors for each rung (see the illustration on the left side of Figure 17). We are not assuming that the rungs are chemically different from one another. We are only assuming that the rungs are chemically different from the sides of the ladder. The colors just help us keep track of each rung, but otherwise have no meaning. Observe that the sides of the Möbius ladder have the form of a folded over figure eight. We assume the molecule is completely flexible, and we deform the sides of the ladder to a planar circle. The result of this deformation is illustrated on the right side of Figure 17. This gives us a nice picture of the Möbius ladder as a circle with three arcs attached at different levels.

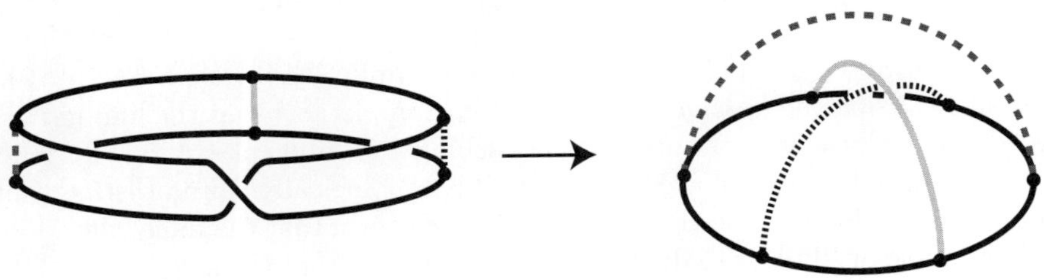

FIGURE 17. We color the rungs of a Möbius ladder and then deform the sides of the ladder into a planar circle.

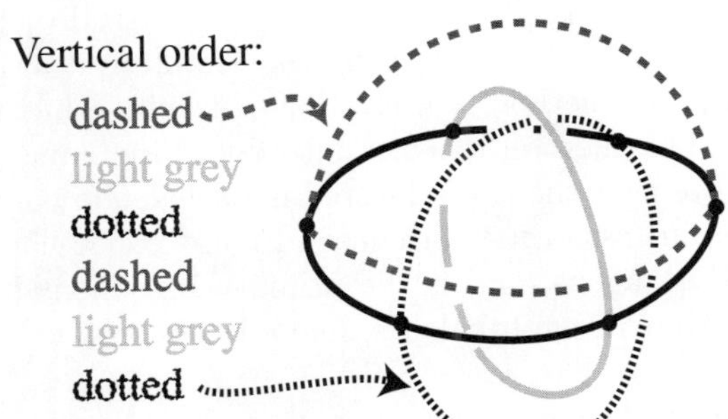

FIGURE 18. The 2-fold branched cover of a Möbius ladder, branched over the sides of the ladder. The vertical descending order of the rungs is: dashed, light grey, dotted, dashed, light grey, dotted.

The intuitive idea of the 2-fold cover of the Möbius ladder branched over the black circle in Figure 17, is illustrated by gluing two copies of the Möbius ladder together along the black circle. This gives us only one black circle, but two sets of rungs stacked one on top of the other. We illustrate this branched cover in Figure 18. In Figure 17, there is a grey dashed rung at the top, a light grey rung in the middle, and a dotted rung at the bottom. In the branched cover this pattern repeats with the endpoints of a rung of a given color attached to the endpoints of the second rung of that same color to create a circle of that color.

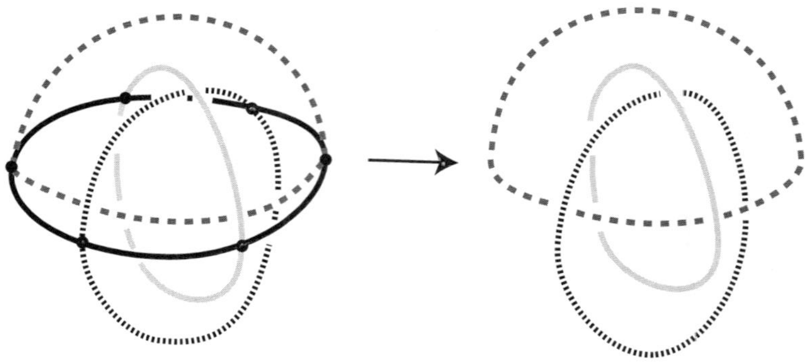

FIGURE 19. If we remove the black circle, we obtain a link.

If we remove the black circle, then we obtain a link (illustrated in Figure 19). It can be shown using the *linking number* (see for example, [**8**]) that the link in Figure 19 is topologically chiral (without distinguishing the different colors). It follows that the branched cover of the Möbius ladder is topologically chiral, distinguishing the link from the black circle. From this, it can be shown that the molecular Möbius ladder is topologically chiral, distinguishing the rungs from the sides.

This is a powerful method to use when a molecule has a special circle (like the sides of the Möbius ladder) which, for chemical reasons, must be invariant under any deformation.

5.3. Using a known topologically chiral molecule. The idea of this method is that if we know a given molecule is topologically chiral, then we use that molecule to show that other molecules which contain it are also topologically chiral. In particular, this method is useful for molecules which contain a Möbius ladder within them. This method was first introduced by Liang and Mislow [**16**] in 1994, in order to prove that the triple layered naphthalenophane molecule is topologically chiral. We illustrate this molecule in Figure 20. A naphthalene molecule consists of a pair of hexagons, and is the primary ingredient in mothballs. This molecule has three layers of naphthalenes, like a triple layered fudge cake made of mothballs.

We consider this molecule as a graph where each corner is a vertex. It is not hard to show that this graph contains a unique longest cycle C, which we highlight in Figure 21. Any deformation of triple layered naphthalenophane to its mirror image would take the cycle C to the corresponding unique longest cycle in the mirror image molecule. Now observe that there are exactly three edges in the graph which have both of their endpoints on C. These edges are indicated as dotted segments in Figure 22. Any deformation of triple layered naphthalenophane to its mirror image would take these three dotted edges to three corresponding edges in the mirror image. Thus if triple layered naphthalenophane could be deformed to its mirror image, then so could the colored graph illustrated in Figure 23. However,

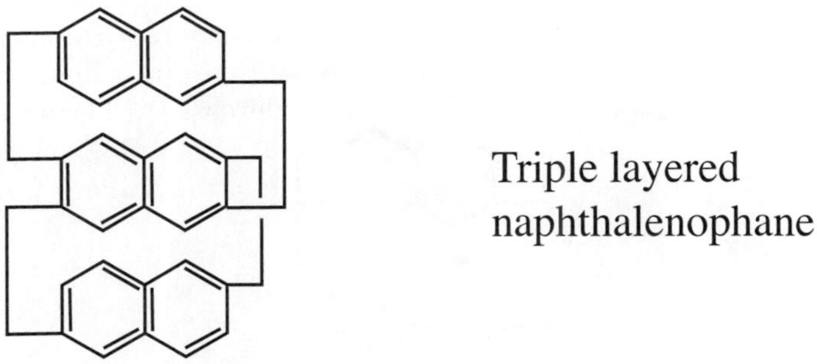

FIGURE 20. This molecule has three levels, each consisting of a pair of hexagons known as a *naphthalene*.

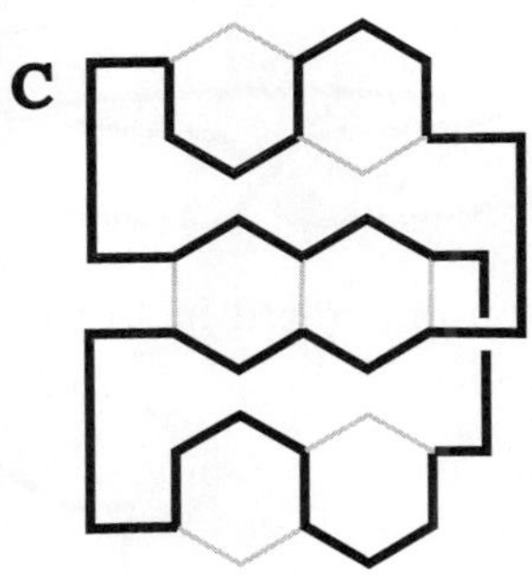

FIGURE 21. Triple layered naphthalenophane contains a unique longest cycle C.

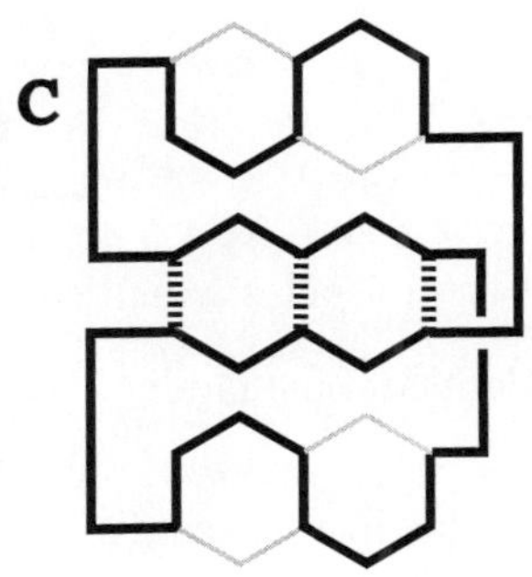

FIGURE 22. The three dotted edges are the only edges in this graph which have both of their endpoints on C.

the graph illustrated in Figure 23 can be deformed to the colored Möbius ladder which is illustrated in Figure 24. Since we know that this colored Möbius ladder is topologically chiral, triple layered napthalenophane must be topologically chiral as well.

This method can sometimes be used even if the molecular graph contains more than one Möbius ladder. For example, consider the graph illustrated in Figure 25. It contains three hexagons at the top and two hexagons at each of the other levels. Thus it contains two equal length longest cycles, and hence two equivalent Möbius

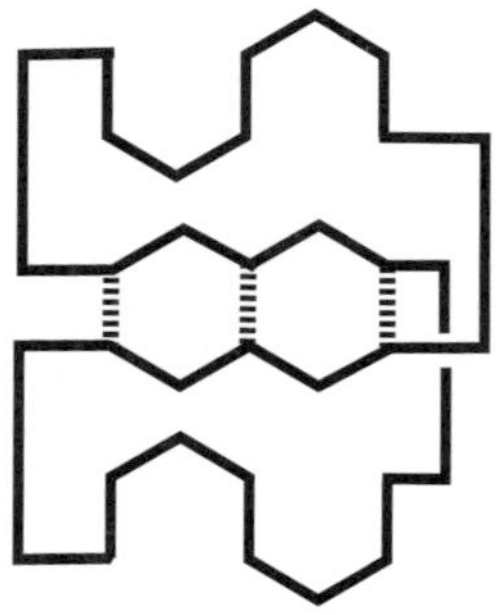

FIGURE 23. If triple layered naphthalenophane could be deformed to its mirror image, then so could this colored graph.

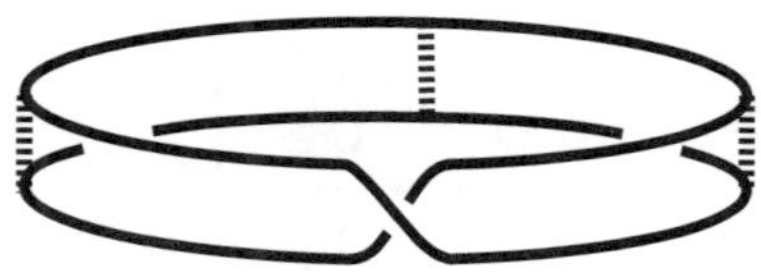

FIGURE 24. The graph illustrated in Figure 23 can be deformed to this colored Möbius ladder.

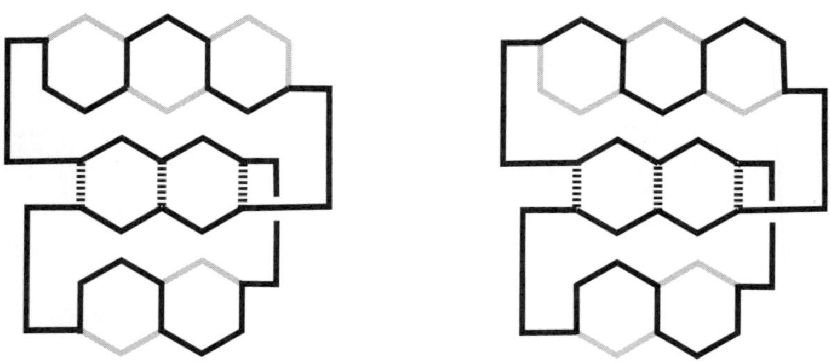

FIGURE 25. This graph contains two equal length longest cycles, and hence two equivalent Möbius ladders, as indicated.

ladders, as illustrated in Figure 25. In this case, a similar argument can be used to show that the graph is topologically chiral.

5.4. A combinatorial approach. The idea of this method is to translate the question of showing that a molecular graph is topologically chiral into an easier question about an abstract graph. We begin with several definitions.

DEFINITION 5.1. An **automorphism** of a graph is a permutation of the vertices of the graph that takes adjacent vertices to adjacent vertices. For a molecular graph, we also require that an automorphism take atoms of a given type to atoms of the same type.

Consider the molecular Möbius ladder illustrated in Figure 16. An automorphism must take oxygens to oxygens and carbons to carbons. Thus an automorphism takes rungs to rungs and sides to sides.

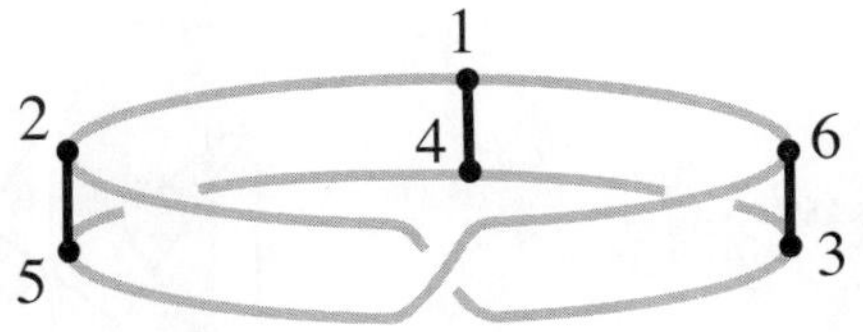

FIGURE 26. The premutation (26)(35) is an automorphism, but the permutation (12) is not an automorphism.

Now consider the Möbius ladder with numbered vertices illustrated in Figure 26. The numbers are to help us keep track of what happens to the vertices during a particular permutation. Observe that the permutation (26)(35) (which interchanges vertices 2 and 6 and interchanges vertices 3 and 5) is an automorphism of the Möbius ladder because it preserves adjacency of vertices. By contrast, the permutation (12) is not an automorphism, since vertex 2 is adjacent to vertex 5, but vertex 1 is not adjacent to vertex 5.

DEFINITION 5.2. The **order** of an automorphism is the smallest number of times you must to perform the automorphism to get every vertex back to its original position.

For example, the order of the automorphism (26)(35) in Figure 26 is 2.

DEFINITION 5.3. The **valence** of a vertex in a graph is the number of edges that contain it.

All of the vertices in the Möbius ladder in Figure 26 have valence 3. By contrast, if we look at the graph of the molecular Möbius ladder in Figure 16 the oxygen atoms have valence 2.

DEFINITION 5.4. The **distance** between two vertices in a graph is the fewest number of edges contained in a path from one to the other.

It is not hard to prove the following two properties of automorphisms that we will use in our subsequent examples.

- An automorphism takes vertices of a particular valence to vertices of the same valence.
- An automorphism takes a pair of vertices which are a certain distance apart in the graph, to a pair of vertices which are that same distance apart.

Finally, we will need to refer to the two special graphs that are illustrated in Figure 27. The complete graph on five vertices, K_5, has an edge between every pair of vertices. The complete bipartite graph on two sets of three vertices, $K_{3,3}$, has an edge joining every vertex in one set to every vertex in the other set, and no edges joining a vertex in one set to another vertex in that same set. The graphs K_5 and $K_{3,3}$ are important because a graph can be embedded in the plane if and only if it does not contain one of these two graphs.

THEOREM 5.5 (Flapan [**9**]). *If a graph contains either K_5 or $K_{3,3}$ and has no order 2 automorphism, then any embedding of the graph in $\mathbb{R}^3$ is topologically chiral.*

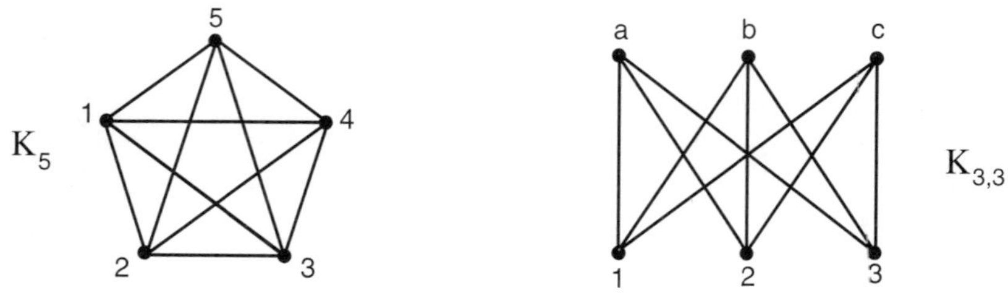

FIGURE 27. A graph has a planar embedding if and only if it does not contain one of these two graphs.

Theorem 5.5 translates a problem about the topology of how graphs are embedded in $\mathbb{R}^3$ into a problem about abstract graphs. All of the topology of Theorem 5.5 is contained its proof, which uses a lot of topological machinery, including: Jaco-Shalen [**13**] and Johannson's [**14**] Characteristic Submanifold Theorem, Thurston's Hyperbolization Theorem [**22**], and Mostow's Rigidity Theorem [**18**].

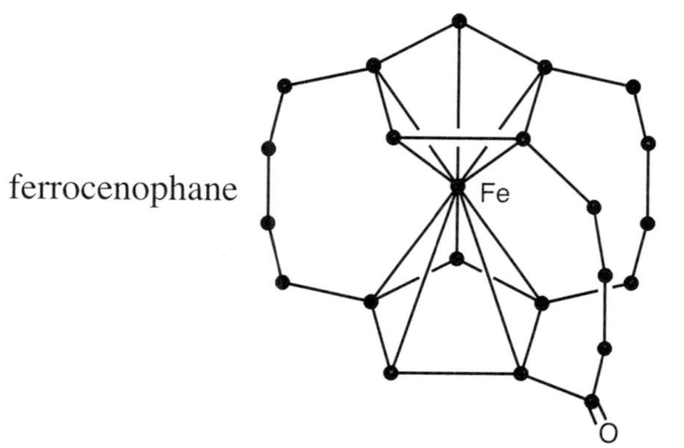

FIGURE 28. We consider the valence of adjacent vertices to argue that Ferrocenophane has no non-trivial automorphisms.

We can use Theorem 5.5 to prove that various molecules are topologically chiral. For example, we consider the molecule ferrocenophane, which is illustrated in Figure 28. All of the atoms in this molecule are carbons except for the iron atom in the center and the oxygen at the bottom. Since an automorphism must take atoms of a given type to atoms of the same type, any automorphism of this graph fixes the single oxygen atom and the single iron atom. Since the oxygen is fixed, the carbon atom attached to the oxygen must also be fixed by any automorphism. Now adjacent atoms must remain adjacent and valence must be preserved. It follows that the two atoms which are adjacent to this carbon are also each fixed by any automorphism. By progressively using the valence of adjacent vertices, we see that any automorphism of ferrocenophane fixes every single atom. Thus ferrocenophane has no non-trivial automorphisms.

It is not apparent at first glance that ferrocenophane contains one of our special graphs, K_5 or $K_{3,3}$. However, Figure 29 illustrates a K_5 that is contained in ferrocenophane. Thus we can apply Theorem 5.5 to conclude that the molecule ferrocenophane is topologically chiral.

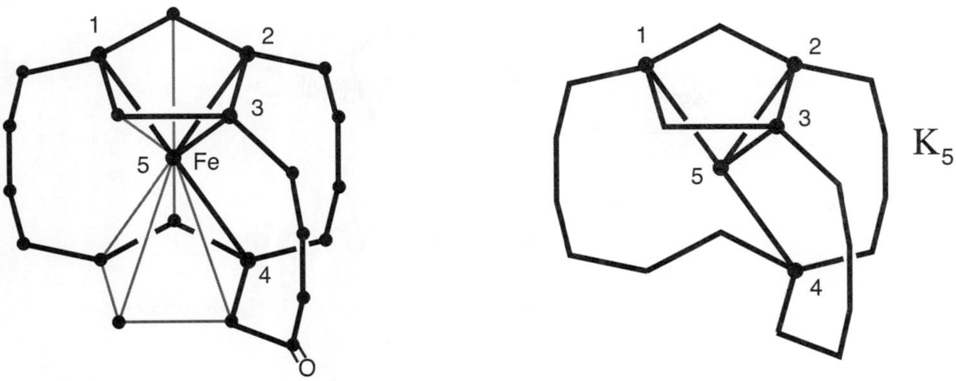

FIGURE 29. Ferrocenophane contains the graph K_5. We illustrate this K_5 without the rest of the graph on the right.

Next consider the Simmons-Paquette molecule, which is illustrated in Figure 30. In this case, we can see that the molecule contains the graph K_5 by simply omitting the vertices of valence two.

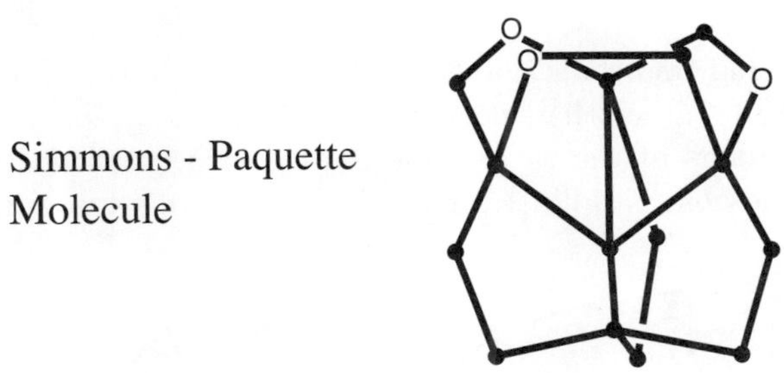

FIGURE 30. To see that the Simmons-Paquette molecule contains K_5 we omit the vertices of valence two.

To prove that the Simmons-Paquette molecule has no automorphisms of order two, we first observe that there is a unique cycle in the graph containing all three oxygen atoms. We label this unique cycle as C in Figure 31. Assume the Simmons-Paquette molecule has an automorphism ϕ of order two. Then ϕ must take the cycle C to itself. Since C has three oxygen atoms and ϕ has order two, one oxygen atom must be fixed by ϕ. Since ϕ preserves valence and adjacency, it now follows that every vertex on C is fixed by ϕ. Starting with the vertices on C, we progressively see that ϕ actually fixes every vertex on the Simmons-Paquette molecule. This contradicts our assumption that ϕ had order two. Thus the Simmons-Paquette molecule cannot have an automorphism of order two. We can now apply Theorem 5.5 to conclude that the Simmons-Paquette molecule is topologically chiral.

Observe that when we use Theorem 5.5 to show that a graph is topologically chiral, then any other embedding of the graph is also topologically chiral. For example, since Theorem 5.5 applies to the Simmons-Paquette molecule, we know that no matter how we embed the graph of the Simmons-Paquette molecule in $\mathbb{R}^3$, it will still be topologically chiral. So we could twist together the edges, or even

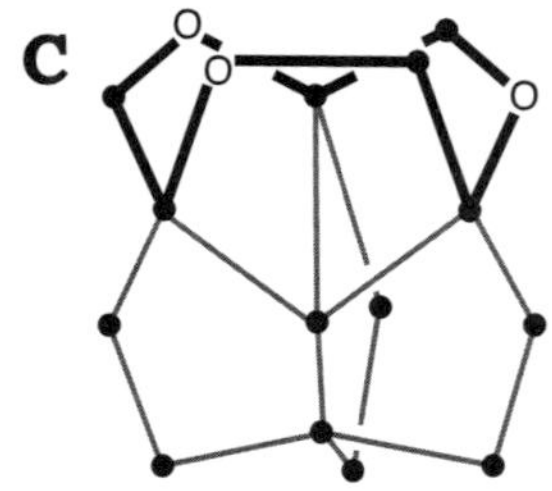

FIGURE 31. C is the unique cycle containing all three oxygens.

tie little knots in some edges, and the embedded graph would still be topologically chiral. Of course, not every embedding of a graph in $\mathbb{R}^3$ represents an actual molecule. In particular, an edge of the Simmons-Paquette molecule represents a bond, and hence is too short to contain a little knot in it. The fact that Theorem 5.5 can be applied to any embedding of a graph motivates us to make the following definition.

DEFINITION 5.6. A graph G is said to be **intrinsically chiral**, if every embedding of G in $\mathbb{R}^3$ is topologically chiral.

This terminology indicates that the topological chirality of the graph is *intrinsic* to the graph itself, rather than depending on how the graph is embedded in $\mathbb{R}^3$. Different embeddings of the same molecule in $\mathbb{R}^3$ are called *stereoisomers*. For example, the two molecules illustrated in Figure 32 are stereoisomers.

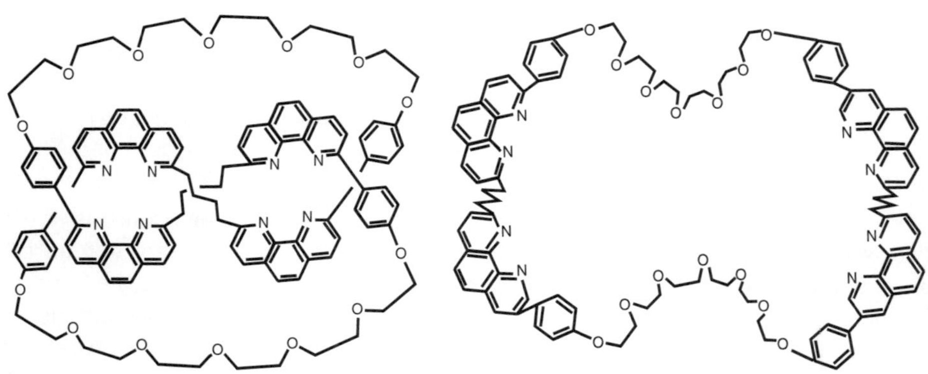

FIGURE 32. The molecule on the left is topologically chiral. The molecule on the right is a stereoisomer that is topologically achiral.

If a molecule is known to be intrinsically chiral, then it and all of its stereoisomers are topologically chiral. Thus we know that all of the stereoisomers of ferrocenophane and the Simmons-Paquette molecule are topologically chiral. Not every topologically chiral molecule is intrinsically chiral. For example, the molecules illustrated in Figure 32 are stereoisomers, but the one on the left is topologically chiral and the one on the right is topologically achiral.

6. Molecular symmetry groups

A molecule which is chemically achiral can be said to have *mirror image symmetry.* In order to predict molecular behavior it is important to understand all of the symmetries of a molecule, not just mirror image symmetry. If we consider only rigid molecular symmetries, then the possible symmetries of a molecule are rotations, reflections, and combinations of rotations and reflections. In more mathematical language we would refer to these rigid symmetries as *isometries* of $\mathbb{R}^3$ which take the molecule to itself. If a molecule is geometrically chiral, then its only rigid symmetries are rotations. The rigid symmetries of a molecule form a group. (We warn the reader that in this section we assume that the reader has some familiarity with group theory.)

DEFINITION 6.1. The **point group** of a molecule is the group of rigid symmetries of the molecule.

This group is called the *point* group because it leaves the *point* at the center of the figure fixed. The point group of the Simmons-Paquette molecule (see Figure 30) is the cyclic group of order three, $\mathbb{Z}_3$, because its symmetries consist of the identity element (i.e., a rotation by 0°), and rotations by 120°, and 240°.

We can also consider the symmetries of an abstract graph independent of any embedding of it in space. The *symmetries* of an abstract graph are just the automorphisms of the graph (which we defined in the last section). As with the rigid symmetries, the automorphisms of a graph form a group.

DEFINITION 6.2. Let Γ be a graph. The **automorphism group of** Γ is the group of automorphisms of Γ.

Every rigid symmetry of a molecule induces an automorphism of its graph. The point group can be associated with this induced subgroup of the automorphism group of the graph. In general, this subgroup is not equal to the entire automorphism group of the graph.

For example, consider the colored Möbius ladder with numbered vertices (illustrated in Figure 26) which represents the molecular Möbius ladder. We have seen that any automorphism of this graph must take rungs to rungs and sides to sides. Observe that the only non-trivial rigid symmetry of this graph is the order two automorphism (23)(56)(14) that is induced by turning the entire graph over left to right. Hence the point group of the molecular Möbius ladder is the cyclic group of order two, $\mathbb{Z}_2$. However, the abstract graph also has an order six automorphism (123456) which is not induced by a rigid symmetry. It follows that the automorphism group is the dihedral group with 12 elements denoted by D_6. Thus, in this case, the point group is associated with a proper subgroup of the automorphism group of the graph.

Next we consider the molecule illustrated in Figure 33. The only non-trivial rigid symmetry of this molecule is a reflection through the plane containing the three hexagons. Thus the point group of the molecule is the cyclic group $\mathbb{Z}_2$. However, the molecule in Figure 33 has a propeller at the top that rotates by 120°. The motion of the propeller is not a rotation of the entire graph; and hence it is not a rigid symmetry of the graph. Thus the point group does not include this symmetry. On the other hand, this motion of the propeller does correspond to an order three automorphism of the graph. Observe that the order two automorphism induced by the reflection does not commute with the order three automorphism induced by the motion of the propeller. Thus, the automorphism group of the graph is D_3, the dihedral group with six elements.

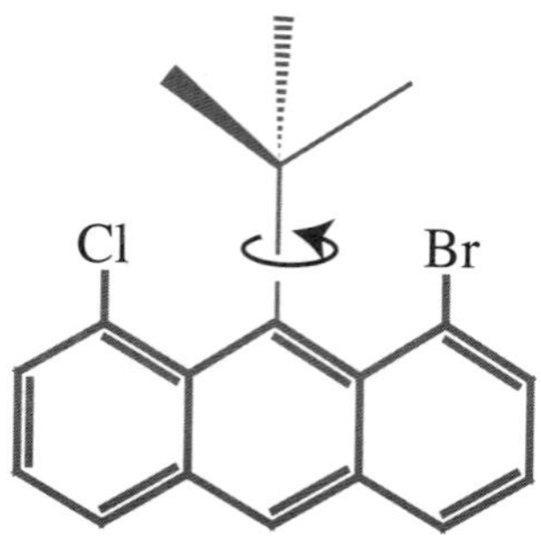

FIGURE 33. The point group of this molecule is $\mathbb{Z}_2$.

Recall from our discussion of rubber gloves that by treating all molecules as rigid, we do not detect that certain flexible, or even partially flexible, molecules are chemically achiral. Examples of this type motivated us to introduce the definition of topological chirality, which treats all molecules as flexible. While there is no mathematical definition of chirality that precisely corresponds to chemical chirality, we saw that chemical chirality falls somewhere between the notions of geometric chirality and topological chirality, depending on the rigidity of the particular molecule involved. Our situation in studying molecular symmetries is analogous. Treating all molecules as rigid, ignores symmetries of a propeller (like that of Figure 33), as well as symmetries which may require even greater flexibility of the molecule. We would like to consider what we would mean be a molecular symmetry group if we treated all molecules as flexible.

DEFINITION 6.3. Let Γ be a graph embedded in $\mathbb{R}^3$. The **topological symmetry group** of Γ is the subgroup of the automorphism group which is induced by deformations and/or reflections of $\mathbb{R}^3$ which take Γ to itself.

The notion of the topological symmetry group of a molecule was first introduced by Jon Simon [**20**] in 1987. Since any rigid symmetry of a graph in $\mathbb{R}^3$ is induced by deformations and/or reflections, the group of automorphisms induced by the point group is a subgroup of the topological symmetry group.

As examples, we determine the topological symmetry groups of some molecules that we have already seen. First we consider the Simmons-Paquette molecule, illustrated in Figure 30. Observe that the only automorphisms of the Simmons-Paquette molecule are induced by rotations of 120° and 240°. Thus the topological symmetry group, the point group, and the automorphism group of the Simmons-Paquette molecule are all equal to $\mathbb{Z}_3$.

Now, we consider the molecular graph illustrated in Figure 33. The topological symmetries of this graph are induced by a rotation of the propeller by 120° and a reflection through the plane containing the hexagons. Both of these symmetries are induced by deformations and/or reflections. Thus, the topological symmetry group is equal to the automorphism group, which we saw above was D_3, the dihedral group with six elements.

Next, we consider the colored Möbius ladder with numbered vertices (illustrated in Figure 26). We saw above that the automorphism group of this graph is D_6, generated by the order two automorphism (23)(56)(14) together with the order six automorphism (123456). The automorphism (23)(56)(14) is induced by turning the graph over left to right, and the automorphism (123456) is induced by rotating the graph by 120° around a central axis, while bringing the half-twist back to its original position. Thus again the topological symmetry group is equal to the automorphism group.

In general, the topological symmetry group of an embedded graph does not have to equal the automorphism group of the graph. Consider the embedded graph illustrated in Figure 34. The automorphism group of this graph is generated by the three involutions (24), (68), and (15)(26)(37)(48). Since these involutions commute, the automorphism group is $\mathbb{Z}_2 \times \mathbb{Z}_2 \times \mathbb{Z}_2$. The involution (24) is induced by turning the left side of the graph over, and the involution (68) is induced by turning the right side of the graph over. However, since it is impossible to deform an unknotted circle into a knotted circle, there is no deformation and/or reflection which induces the involution (15)(26)(37)(48). Thus the topological symmetry group of this embedded graph is $\mathbb{Z}_2 \times \mathbb{Z}_2$, which is a proper subgroup of the automorphism group of the graph.

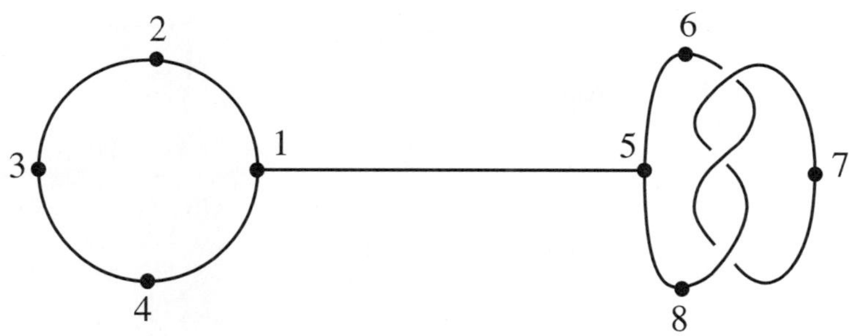

FIGURE 34. The topological symmetry group of this embedded graph is not equal to the automorphism group.

In 1938, Frucht [**12**] proved that every finite group is equal to the automorphism group of some graph. It is natural to wonder whether every finite group is equal to the topological symmetry group of some embedded graph. In 2005, we answered this question in the negative by proving Theorem 6.4.

THEOREM 6.4 (Flapan, Naimi, Pommersheim, Tamvakis [**10**]). *There is no graph Γ embedded in $\mathbb{R}^3$ such that the topological symmetry group of Γ is equal to an alternating group A_n with $n > 5$.*

In order to compare the different approaches to understanding symmetries of molecules, we identify the point group with the subgroup of the automorphism group of its graph, and we make the following final definition.

DEFINITION 6.5. The **chemical symmetry group** of a molecule is the subgroup of the automorphism group which is induced by chemically achievable motions taking the molecule to itself or to its mirror image.

A rotation of a molecule is a chemically achievable motion. If an automorphism is induced by a reflection of the molecule, then the molecule and its mirror image are identical as rigid objects. Hence there is a chemically achievable motion taking the molecule to its mirror image. Thus the point group is a subgroup of the chemical symmetry group. Every chemically achievable motion is a deformation. If a chemically achievable motion takes the molecule to its mirror image, then this motion followed by a reflection takes the molecule to itself. Thus every automorphism which is the result of a chemically achievable motion taking the molecule to itself or to its mirror image is also the result of a deformation and/or a reflection. So the chemical symmetry group is a subgroup of the topological symmetry group. Finally, every automorphism which is induced by a deformation and/or a reflection is certainly an element of the automorphism group of the graph. Thus the topological symmetry group is a subgroup of the automorphism group of a graph. The relationship between the different symmetry groups is illustrated in Figure 35. As we have seen above, these subgroups may or may not be proper subgroups depending on the specific example involved.

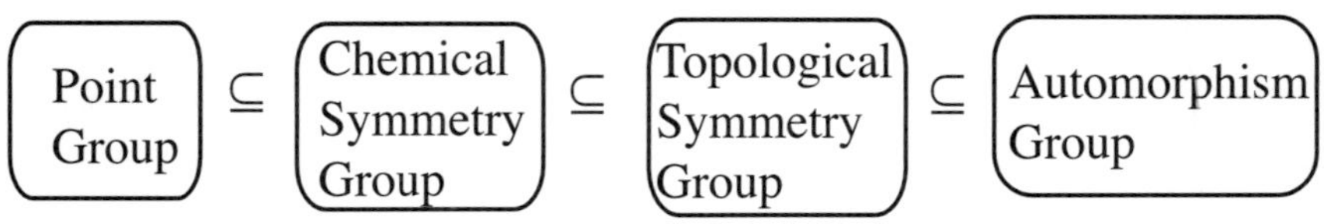

FIGURE 35. The relationship between the different types of symmetry groups.

For a more detailed presentation of topological chirality and symmetries of non-rigid molecules, the interested reader should see [**8**].

References

[1] J. W. Alexander, *Topological Invariants of Knots and Links*, Trans. Amer. Math. Soc. **30** (1928), 275-306.

[2] M. Boileau, B. Leeb, and J. Porti, *Uniformization of small 3-orbifolds*, C. R. Acad. Sci. Paris Sér. I Math. **332** (2001), 57–62.

[3] M. Boileau and J. Porti, *Geometrization of 3-orbifolds of cyclic type*, Astérisque **272**, SMF Publications, Paris, 2001.

[4] D. Cooper, C. Hodgson, and S. Kerckhoff, *Three-dimensional orbifolds and cone-manifolds*, Mathematical Society of Japan Memoirs **5**, Tokyo, 2000.

[5] J.-C. Chambron, J.-P. Sauvage, K. Mislow, *A chemically achiral molecule with no rigid achiral presentations* J. Am. Chem. Soc. **114** (1997), 9558-9559.

[6] J. Conway, *An enumeration of knots and links and some of their related properties* Computational problems in Abstract Algebra, Proc. Conf. Oxford, Pergammon, Oxford (1967) 329-358.

[7] C. Dietrich-Buchecker, J.-P. Sauvage, *A synthetic molecular trefoil knot*, Angew. Chem. Int. Ed. Engl., **28** (1989), 189-192.

[8] E. Flapan, *When Topology Meets Chemistry*, Cambridge University Press and the Mathematical Association of America (2000).

[9] E. Flapan, *Rigidity of Graph Symmetries in the 3-sphere*, Journal of Knot Theory and its Ramifications, **4** (1994,) 373-388.

[10] E. Flapan, R. Naimi, J. Pommersheim, and H. Tamvakis, *Topological Symmetry groups of graphs embedded in the 3-sphere* Comment. Math. Helv., **80** (2005), 317-354.

[11] E. Flapan, R. Naimi and H. Tamvakis, *Topological symmetry groups of complete graphs in the 3-sphere*, J. London Math. Soc., **73** (2006), 237-251.

[12] R. Frucht, *Herstellung von graphen mit vorgegebener abstrakter Gruppe*, Compositio Math. **6** (1938), 239–250.
[13] W. Jaco and P. Shalen, *Seifert fibred spaces in 3-manifolds*, Memoirs Amer. Math. Soc., **220** Amer. Math. Soc., Providence, 1979.
[14] K. Johannson, *Homotopy equivalences of 3–manifolds with boundaries*, Lecture Notes in Mathematics, **761**, Springer-Verlag, New York, Berlin, Heidelberg, 1979.
[15] V. F. R. Jones, *A polynomial invariant for knots via Neumann algebras*, Bull. Amer. Math. Soc. **89** (1985), 103-111.
[16] C. Liang and K. Mislow, *Classification of topologically chiral molecules*, J. of Mathematical Chemistry, **15** (1994), 245-260.
[17] K. Mislow and R. Bolstad, *Molecular dissymmetry and optical inactivity*, J. Am. Chem. Soc., **77** (1955), 6712-6713.
[18] G. Mostow, *Strong Rigidity for Locally Symmetric Spaces*, Ann. Math. Studies, **78**, Princeton University Press, Princeton, (1973).
[19] J. Simon, *Topological chirality of certain molecules*, Topology, **25** (1986), 229-235.
[20] J. Simon, *A topological approach to the stereochemistry of nonrigid molecules*, Graph theory and topology in chemistry (Athens, GA, 1987), Stud. Phys. Theoret. Chem. **51**, 43–75, Elsevier, Amsterdam, 1987.
[21] P. A. Smith, *Transformations of finite period II*, Annals of Math. **40** (1939), 690–711.
[22] W. Thurston, *Three-dimensional manifolds, Kleinian groups and hyperbolic geometry*, Bull. Amer. Math. Soc. **6** (1982), 357–381.
[23] N. Van Gulick, *Theoretical aspects of the linked ring problem*, New J. Chem. **17** (1993), 645-653.
[24] D. Walba, R. Richards, R.C. Haltiwanger, *Total synthesis of the first molecular Möbius Strip*, J. Amer. Chem. Soc.**104** (1982), 3219-3221.

Mathematics Department, Pomona College, Claremont, CA 91711, USA
E-mail address: eflapan@pomona.edu

Proceedings of Symposia in Applied Mathematics
Volume **66**, 2009

DNA Topology

Dorothy Buck

ABSTRACT. This article briefly surveys the topological properties of DNA. We sketch how the helical structure of DNA intimately links the geometry and topology of the molecule. We then discuss the proteins responsible for regulating this DNA geometry and topology, and the mathematical questions that arise when understanding them. We next turn our attention to the site-specific recombinases, proteins that change DNA topology as a by-product of their primary action. We discuss one topological model that predicts the types of knots and links that arise from site-specific recombination. We conclude with a discussion of the tangle model, and give an extended example where it was useful in understanding the mechanism of a site-specific recombinase.

1. Introduction

Historically, knot theorists were motivated in part by chemical considerations. Lord Kelvin conjectured that different elements were comprised of different knotted vortices of the then-fashionable pervasive ether. (See [**85**] for a great introduction to the early history of knot theory.) His friend Peter Guthrie Tait, who was also interested in vortex rings, began to classify knots and links, and produced the first modern knot tables—on display at the British Library.

While the original applied motivation for understanding knots evaporated, knot theory is once again being utilized by biologists and chemists. Beginning in the mid-1980s, when linked and knotted DNA was first experimentally found, topologists have played an increasing role in exploring the ramifications. This brief article surveys some of the landscape charted out in the new interdisciplinary territory of DNA Topology.

1.1. What this article covers. This article is based on a lecture given at the 2008 AMS Short Course "Applications of Knot Theory", organized by Erica Flapan and myself, and so retains some of the informal tone of a lecture. The audience—then and (intended) now—is mathematicians who know some topology, and are interested in learning more about its interplay with molecular biology. Accordingly, here we give an introductory overview of this, focussing on knots in a fundamental biochemical setting—those formed by DNA. We begin in Section 2, by sketching the basic primary and secondary structure of DNA, and then shift our

2000 *Mathematics Subject Classification.* Primary 92C40, 92E10, 57M25.
Key words and phrases. Knot Theory, Molecular Biology.

focus to the more topologically relevant tertiary structure. In Section 3, we model DNA as a topological ribbon. We describe the Serret-Frenet framing, and introduce the fundamental relation $Lk = Tw + Wr$. In Section 4, we discuss the importance of DNA supercoiling for compactification, base-pair access and the free energy. In Section 5, we give a brief overview of the two families of proteins, type I and type II topoisomerases, that regulate DNA supercoiling and DNA knotting/linking, respectively. We touch on the topological question of type II topoisomerase unknotting, and its relation to the more general subject of knot adjacency. In Section 6, we discuss the primary biological techniques to separate and distinguish DNA knots and links. In Section 7, we introduce another family of DNA topology-changing proteins, the site-specific recombinases. And finally, in Section 8, we discuss the tangle model for site-specific recombination.

1.2. What this article doesn't cover, and where to read more. Both for space and expertise constraints, this article is a necessarily brief introduction to an extraordinarily rich interdisciplinary area. So unfortunately, there are a number of fascinating topics that we've only touched on or entirely neglected. For example, we only briefly sketch the basic background material for the energetics of DNA supercoiling. Similarly, our understanding of the structure and mechanism of type II topoisomerases has radically improved in the past 18 months, particularly with the emergence of the first crystal structure of a yeast-DNA co-complex [**29**]. Interested readers may also want to consult the Proceedings of the Topoisomerase conference held in Norwich in July 2008 for more details. Likewise, we will gloss over the many models of DNA migration through an electrophoretic gel. Also, we will not discuss the beautiful rod models for DNA.

There are a number of general expository treatments of (facets of) this area. We encourage the interested reader to consult, for example, the books of Vologodskii [**98**] and Frank-Kanemetskii [**37**], as well as overview articles by Pohl and Weber (the last in French, but with a nice historical overview), and the similar 1992 AMS short course [**28**]. From a more biological viewpoint, we recommend Maxwell and Bates' *DNA Topology* [**65**], and *Mobile DNA* for more on recombination. [**21**].

2. Structure of DNA

2.1. Primary and secondary structure. DNA, Deoxyribonucleic Acid, is the molecule (or group of molecules) responsible for encoding all genetic information and instructions in living organisms. The information stored in DNA determines all hereditary traits, for example, eye colour and susceptibility to inherited diseases. For this reason, DNA is often referred to as the blueprint (or more accurately the recipe book) for life.

The structure of DNA was famously pieced together by Francis Crick and James Watson, based on beautiful X-ray images of Rosalind Franklin [**102**]. DNA is composed of pairs of repeated units called nucleotides. Each nucleotide consists of a phosphate group, a ($2'$-deoxyribose) sugar, and one of four bases: Adenine, Thymine, Guanine or Cytosine.

These nucleotides form a twisted ladder, the famous 'double helix' [**102**]. The alternating sugar and phosphate groups form the backbones, or sides of the ladder, and stacked pairs of bases form the rungs of the ladder. These bases always pair either A with T or G with C. An AT 'rung' is formed via 2 hydrogen bonds via A and T, while the stronger GC rung consists of 3 hydrogen bonds between G and C.

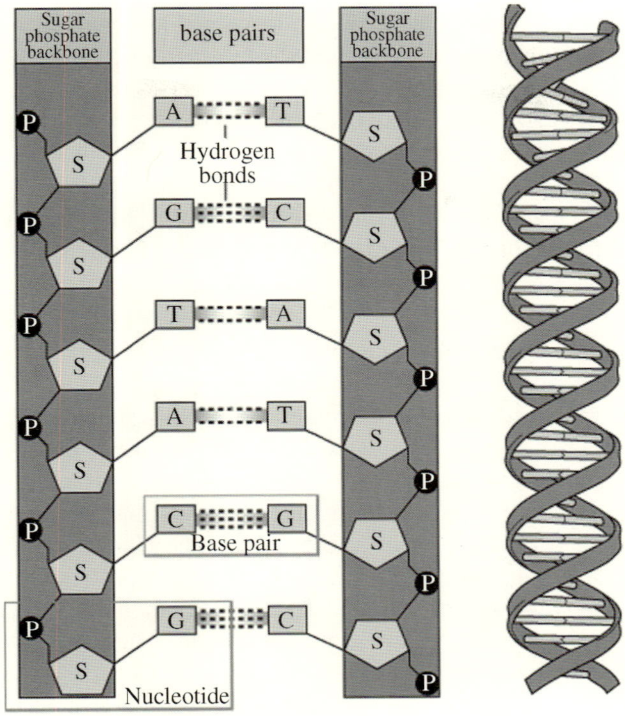

FIGURE 1. Primary and secondary structure of DNA.

The length of DNA is measured in terms of these base pairs (bp); for example, human genomic DNA is approximately 3 billion base pairs, while that of the most common bacteria, *E. coli*, is approximately 4.4 million base pairs. In more absolute measurements, each nucleotide is approximately 3 Angstroms ($= 3 \times 10^{-10}$m) in length. The average width of the double helix is approximately 25 Angstroms.

The sugar in DNA is a pentose sugar, with five carbons. Each sugar shares a (phosphodiester) bond with its two adjacent phosphates, one at its $3'$ hydroxyl (OH-group) and one at its $5'$ hydroxyl. These (phosphodiester) bonds then give each backbone an orientation, and the two strands are aligned antiparallel: the so-called Crick strand running $5'$ to $3'$ and the Watson strand running $3'$ to $5'$. Because of the monogamous base pairing, DNA sequences are typically written with respect to the $5'$ to $3'$ direction, (*i.e.* along the Crick backbone), for example, GATTACA.

The precise spelling of this sequence, the primary structure, encodes all hereditary information, as well as instructions for all cellular processes.

In vitro, the zoo of DNA forms is staggeringly diverse; for example, see Ned Seeman's companion article for some fascinating nanostructures, including truncated octrahedra and knots, synthesized from short (typically single-stranded) sequences [**82**]. Even *in vivo*, there is a wide menagerie—including left-handed helices, right-handed helices whose backbones zigzag, helices whose bases are on the outside of the sugar-phosphate backbone, or even short stem and loops. (See e.g. Maxwell and Bates [**65**] for a description of these forms.) In this article, we will focus our attention on the most prevalent (and hence relevant) structure, B-form DNA. This is a regular right-handed helix with bases on the inside. Depending on its environment, there are approximately 10.5 base pairs per complete turn of the helix. We will discuss this in more detail below. There is both a major and minor groove in the helix. These grooves allow access to the base pairs.

The secondary structure of DNA, its helical geometry, also has deep biological consequences. For example, the double helix is more difficult to unzip than a straight ladder. This difficulty in accessing the base pair rungs results in fewer accidental changes to the DNA sequence, thus protecting the genetic code.

2.2. Tertiary structure. A decade after the double helix discovery, Vinograd found the fundamental tertiary structural feature: the axis of the DNA can also be coiled in space [**84**], leading to 'supercoiled' DNA. (To visualize this, imagine the handset cord of an often-used phone: it is highly twisted, possibly crossing itself and is under tension.)

The DNA axis is often linear, as in the case of human genomic DNA. Linear DNA is often attached to a protein scaffolding which constrains the topology, for example during nucleosomal compactification discussed in Section 4.1 below.

Perhaps more interestingly, the DNA axis is frequently circular—this occurs when the ends of the 2 backbone strands are covalently bonded. Bacterial genomic DNA, chloroplast DNA, and human mitochondrial DNA are all examples of DNA with a circular axis. Additionally, most DNA used in biochemical labs, plasmid DNA, is circular.

Even more interestingly, the central DNA axis can also be knotted or linked. For example, DNA knots and links (known as *catenanes* in the biological literature) occur during replication (DNA copying) and recombination (DNA rearranging). We will discuss DNA knots and links below in more detail. For now, we emphasize that the constrained or circular axis, coupled with the two levels of coiling (the double helix and supercoiling), already leads to the nontrivial topology of DNA.

3. DNA as a topological ribbon: $Lk = Tw + Wr$. The interplay between topology and geometry

From the outset, characterizing supercoiled DNA required a combination of mathematicians and biologists. Călugăreanu [**13**] and White [**105**] formulated a fundamental relation for space curves: $Lk = Tw + Wr$. Two geometric quantities, *Twist* and *Writhe* (that may change under deformations of the curves) sum to a topological quantity, *Linking Number* (which is invariant under such deformations). It took the combined efforts of Fuller (a mathematician) and Vinograd (a biologist) to translate Călugăreanu and White's and relation into biological terms and recognize its fundamental importance for DNA [**39**].

To understand this relation, we shall model a molecule of circular DNA as a twisted ribbon. For each of the two backbones, we associate a curve in 3-space (an edge of the ribbon). By convention we name the backbones C and W, after Crick and Watson. The orientation of these two ribbon edges will be taken to be parallel, and so not actually inheriting the natural orientation from the underlying chemical structure.

These curves C and W are linked around each other, and in fact the ribbon itself can be knotted. Likewise, if we consider more than one DNA molecule, the ribbons themselves can be linked. We formalize this in terms of the linking number, Lk:

DEFINITION 3.1. Given a projection of an oriented two-component link, to each crossing assign a ± 1, depending on the orientation of the two strands.[1] The *linking number, Lk* is one-half their sum:

$$Lk = \frac{1}{2} \sum_{crossings} \pm 1.$$

As discussed in Colin Adams' companion article, Lk is independent of the particular projection and is a topological invariant of the embedding of the ribbon [**1**]. (As before, we say two embedded ribbons are *equivalent* if there exists an orientation-preserving diffeomorphism of S^3 or $\mathbb{R}^3$ which sends one ribbon onto the other.

For a circular DNA molecule, the two DNA backbones, C and W, form a $(2, m)$-torus link. A $(2, m)$-torus knot or link is one which can be drawn so that all of its crossings occur as a row of m (positive or negative plectonemic) crossings, as illustrated in Figure 10. Note that if m is odd, then $T(2, m)$ is a knot and if m is even, then $T(2, m)$ is a link.

Lk can be defined in a variety of equivalent ways; see Rolfsen [**79**]. One alternate description of Lk is in terms of Gauss' integral:

$$(3.1)\quad Lk = \frac{1}{4\pi} \int_C \int_W \frac{(x'-x)(dydz'-dzdy') + (y'-y)(dzdx'-dxdz') + (z'-z)(dxdy'-dydx')}{[(x'-x)^2 + (y'-y)^2 + (z'-z)^2]^{3/2}}$$

where (x, y, z) ranges over C and (x', y', z') ranges over W. (Note: This is the pullback of the area form on S^2 in S^3.)

3.1. Frenet framing. Roughly, we construct the Frenet framing for a regular closed space curve $\gamma(s)$, as follows. (For a more detailed treatment, see [**74**].) Suppose we have a simple closed curve $\gamma(s)$, parametrized by arc length s. Then the tangent vector, T, has length 1: $||T|| = ||\dot{\gamma}(s)|| = 1$. Then $\dot{\gamma}(s).\dot{\gamma}(s) = 1$, and so $\ddot{\gamma}(s)$ is perpendicular to $\dot{\gamma}(s)$, i.e. $\ddot{\gamma}(s)$ is a normal vector. (Note: Since $\gamma(s)$ is regular, then $\dot{\gamma}(s) \neq 0$ by definition.) For simplicity, we will normalize this so that our normal vector, N, is also a unit vector: $N = \frac{\ddot{\gamma}(s)}{||\ddot{\gamma}(s)||}$ for all s such that $\ddot{\gamma}(s) \neq 0$. We then consider the binormal $B := T \times N$.

Since T, N and B are orthonormal, they form a basis so that any other vector can be described as a linear combination of these—in particular, each of the vectors $\dot{T}, \dot{N}$ and $\dot{B}$ can be described as such. Given $\dot{T} = aT + \kappa N + cB$, for some a, κ, c, then $a = 0$ since by above $\dot{T}$ is perpendicular to T. Also, since $N = \frac{\dot{T}}{||\dot{T}||}$, then $c = 0$ and so $\kappa = ||\dot{T}||$. (κ is often called the curvature; at a given point on $\gamma(s)$, since T and $\dot{T}$ span a plane, within that plane one can draw a circle tangent to $\gamma(s)$ with maximal radius $\frac{1}{\kappa}$.)

Similarly, $\dot{N} = -\kappa T + \tau B$, where τ is often called the torsion, a measure of how nonplanar the curve is. Finally, $\dot{B} = \dot{T} \times N + T \times \dot{N} = 0 + T \times (-\kappa T + \tau B) = -\kappa\tau T + \tau^2 B = \tau \dot{N}$.

[1]Unfortunately, the convention adopted by biologists in the field is opposite to the usual topologists' convention.

These 3 relations form the Serret-Frenet equations:

$$\dot{T} = \kappa N,$$
$$\dot{N} = -\kappa T + \tau B, \tag{3.2}$$
$$\dot{B} = -\tau N.$$

These crossings of the 2 backbone curves come in two flavours: local and nonlocal. Local crossings occur when viewing a segment of the ribbon almost edge-on, while nonlocal crossings occur when one segment of the ribbon crosses entirely over another. In the latter case the axis of the ribbon is also crossing over itself. (For an elegant description of this, and a rigorous characterization of Twist in terms of these local crossings; see [**25**].)

3.2. Twist. The local crossings reflect how tightly the ribbon wraps (or twists) around its central axis. In terms of our DNA double helix, these local crossings indicate the helical pitch (i.e. number of base pairs per complete revolution).

Intuitively, one can imagine stringing a penny onto the curve. If we marked the northern pole, then Tw would measure how this marked point changes as we slide the penny along the curve.

More precisely, we quantify this as *Twist*, Tw, using the Frenet framing, as the total torsion of the curve $\gamma(s)$:

$$Tw = \frac{1}{2\pi}\int_{\gamma(s)} (T \times N).\dot{N} ds = \frac{1}{2\pi}\int_{\gamma(s)} \tau ds.$$

Then in the Frenet framing, $(T \times N).\dot{N}$ is the projection of $\dot{N}$ in the direction of B, so by the Serret-Frenet equations, $= \tau$.

3.3. Writhe. Writhe is an indicator of how the ribbon axis itself is contorted in space. If we consider the axis as inheriting the orientation of the two ribbon edges, we can sum the signed crossings of the axis with itself. This quantity is called the *writhe*, Wr. Note that Wr is projection-dependent, in particular a Reidemeister Type I move will contribute a ± 1 to Wr. Unsurprisingly then, Wr is not a topological invariant—it depends on the particular geometry of a given projection. (Note that this terminology varies in the literature. For example, Weber calls the signed sum of the crossings of a projection the *Tait number* of the projection. Then the writhe of a simple closed curve in S^3 or $\mathbb{R}^3$ is the average of the Tait numbers over all projections [**103**]. However, we will follow the (perhaps misleading) most common nomenclature used in the biological literature.)

More formally, we can consider writhe in terms of Gauss' integral, as in formula (3.1), with the axis representing both the C and W curve.

With this definition Wr can then be thought of in terms of the Gauss integral as Lk above, with both integrations along the same curve (the ribbon axis).

For a DNA molecule, the self-crossings of the DNA axis contribute to writhe. The closest macroscopic analogue is the handset of an old-fashioned telephone receiver, attached to the base by a helical cord. During normal use, the cord gains writhe, which is released by holding the cord near the base and letting the dangling handset spin.

As mentioned above, this relation between the topological and geometric properties of a ribbon was first discovered by Călugăreanu [**13**] and White [**105**]. Its relevance to DNA was first noted by Fuller [**39**].

THEOREM 3.2 (Călugăreanu-White-Fuller). $Lk = Tw + Wr$.

Biologists often visualize this correspondence using a length of Taigon clear plastic tubing, with two parallel curves drawn along the length, each representing one edge of the ribbon. By twisting and/or coiling the tubing before closing the ends, it helps to visualize the relation between Twist and Writhe.

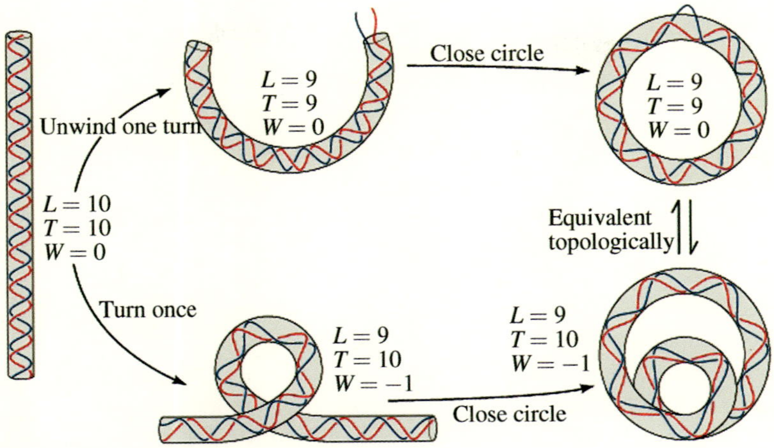

FIGURE 2. Twist and Writhe. From *Biochemistry* by Donald Voet and Judith G. Voet (2004). Reproduced with permission of John Wiley and Sons Ltd.

For a modern treatment of this, see Dennis and Hannay [**25**]. For a more detailed discussion of these differential geometric notions in terms of DNA, see Weber [**103**] or Pohl [**73**].

4. Biological importance of DNA supercoiling

Recall that within the context of circular DNA, nontrivial writhe occurs most commonly when the DNA axis wraps around itself in a helical manner, as discovered by Vinograd and colleagues [**84**]. This is more precisely referred to as *plectonemic supercoiling*, to reflect the coiled (writhed) axis of the already coiled (helical) ribbon. Throughout we will use the term 'supercoiling' to mean 'plectonemic supercoiling'. (DNA can also wrap around proteins, as discussed below, and this type of nontrivial writhe is referred to as *solenoidal supercoiling.*)

This 'supercoiled' DNA was found to be the native state in virtually all cells [**17**]. DNA extracted from cells from a wide variety of organisms, including cauliflower, humans, mice and monkeys, was found to be (plectonemically) supercoiled at approximately the same density.

We can quantify this density as follows. Given N, the number of base pairs of DNA and h, the number of base pairs per helical repeat in given experimental conditions, then the specific linking difference is σ:

$$\sigma = \frac{Lk - N/h}{N/h}.$$

DNA supercoiling is ubiquitous because supercoiling is crucial for three primary reasons. Firstly, it compactifies the DNA molecule. Secondly, the handedness of the supercoiling opens the major/minor grooves for easier access to the base pairs. And thirdly, supercoiled DNA provides an important source of free energy for cellular reactions.

4.1. Compactification. In all organisms, the physical length of genomic DNA is considerably larger than the diameter of the region it resides within. In prokaryotes, the DNA floats freely around the cytoplasm, while in eukaryotes the DNA is encapsulated within the cell nucleus. Eukaryotic genomic DNA is orders of magnitude longer, so while the typical compactification needed to store DNA in prokaryotes is significant, it is even more dramatic in eukaryotes. For example, the most common bacteria *E. coli* has genomic DNA of length approximately 1.5mm, and cell diameter less than 1 μ meter, requiring on the order of 10^3 compactification. Human genomic DNA is approximately a meter in length and a typical cell nucleus has diameter less than 1μm, so the compactification needed is on the order of 10^6. One of the most challenging packing problems is that of the South American lungfish, whose DNA is on the order of 35 meters.

Supercoiling is the first step of a hierarchy of compactification, the later stages of which are still poorly understood. The (plectonemically) supercoiled DNA molecule then wraps around large proteins called histones. Then the histones string together like pearls on a necklace. This chain of histones is then further compacted, possibly as suggested in Figure 3.

Thus supercoiling allows the DNA to be packaged in a tightly confined volume in a highly structured, organized manner. The orderliness of this process is crucial for efficient information retrieval—as (segments of) DNA must be accessed rapidly and constantly by proteins.

4.2. Access to the base pairs. The handedness of DNA supercoils *in vivo* is negative. That is, the helix formed by the DNA axis is opposite in handedness to the original double helix. If the circular molecule remains covalently closed while supercoiling, then the linking number Lk remains constant. Then $Lk = Tw + Wr$ implies that their must be a compensating change in Tw.

The handedness of the supercoils means that the molecule becomes undertwisted. Then the major and minor grooves, illustrated in Figure 1, open a small amount. This allows easier access to the corresponding base pairs, facilitating a wide variety of reactions. It also makes it easier to separate the DNA strands, necessary for example for copying DNA.

4.3. Free energy. As noted above, unzipping the DNA double helix can be difficult. But many essential reactions, including DNA copying, demand exactly this. Fortunately, Theorem 3.2 implies that the linking number of negatively supercoiled DNA (the natural state of DNA in cells) will be lower than that of relaxed DNA. It is this linking number 'deficit' of supercoiled DNA that provides the necessary energy for local strand separation.

In addition, for many biological processes involving DNA, including helical wrapping around histones and phage head packing, the DNA axis must be twisted or bent [**83**]. To understand the energetics of these processes, we briefly discuss how the free energy depends on linking difference.

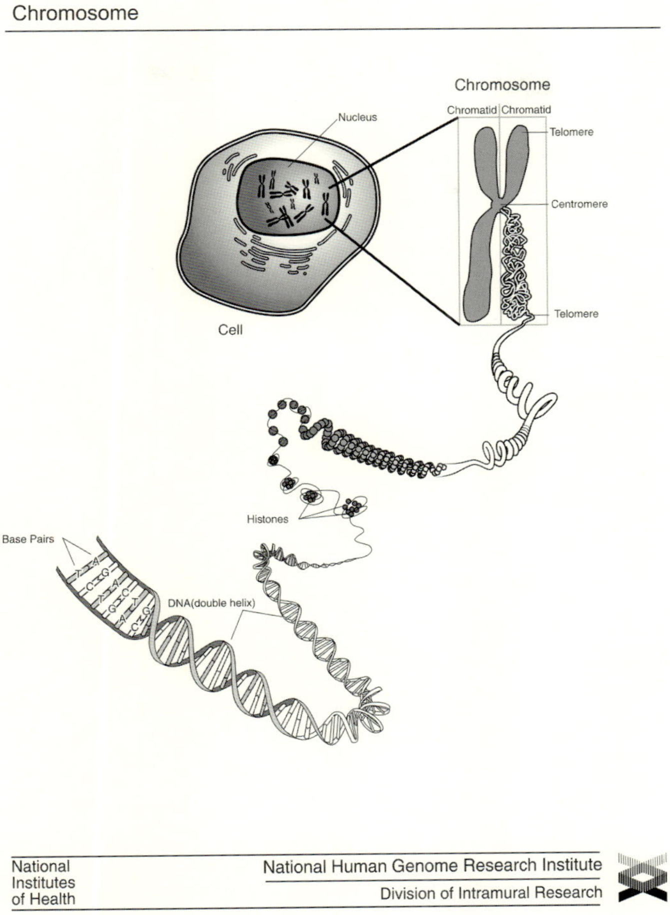

FIGURE 3. Compactifying DNA. Courtesy: National Human Genome Research Institute

For DNA rings larger than 2000bp, it has been experimentally shown in [**27**], [**75**], that the free energy of supercoiling, ΔG, obeys the quadratic relation

$$\Delta G = K(\alpha - \alpha_0)^2 \tag{4.1}$$

where K is a proportionality constant (which depends on N, the number of base pairs), α is the linking number, and α_0 (for large DNA) is the average linking number. (Note that α_0 can be independently defined as N/h_0 where h_0 is the helical repeat, or average number of base pairs over which the helical twist of the DNA double helix increases by one.) For large N, the product NK has been experimentally shown to be independent of length [**49**].

5. Regulating DNA topology

DNA molecules that differ only in Lk (not, *e.g.* sequence or length) are called *topoisomers.* By standard thermodynamics, the concentration of DNA with linking number α_i, $[\alpha_i]$, is

$$[\alpha_i] = \frac{1}{Z} \exp\left(\frac{-\Delta G}{RT}\right), \tag{5.1}$$

where Z is a normalization constant, T is the temperature in Kelvin, and R is Boltzmann's constant. Thus, the concentration of a given topoisomer with linking number α_i, is

$$[\alpha_i] = \frac{1}{Z} \exp\left(\frac{-K(\alpha_i - \alpha_0)^2}{RT}\right). \tag{5.2}$$

The graph of α_i v. $[\alpha_i]$ for equilibrium populations of topoisomers of large DNA rings is a normal distribution with center α_0 and standard deviation $\sqrt{\frac{RT}{2K}}$.

To maintain this equilibrium population, two families of proteins have evolved, the *topoisomerases*[2]. Topoisomerases are so-called because the primary function is to interconvert between topoisomers. Topoisomerases are classified into two families, depending on whether they cleave 1 DNA backbone and hence change the linking number by steps of 1 (Type I topoisomerases) or both backbones and hence change linking number in steps of 2 (Type II topoisomerases).

We briefly discuss these two families below, but emphasize that we are glossing over many of the subtleties of topoisomerases—in particular, the minor differences between family members of a given type. We refer the reader who is interested in exploring more in these remarkable proteins to look at, e.g. [**65**].

5.1. Regulating DNA supercoiling: Type I topoisomerases. Given the fundamental importance of DNA supercoiling, it is perhaps not surprising that there are proteins whose sole function is to regulate the amount of supercoiling—the Topoisomerase I family. Members of this family relax (negative, and in eukaryotes, also positive) supercoils (See Figure 4.)

Type I topoisomerases effectively harness Theorem 3.2, to convert a change in twist to a change in writhe (supercoiling). They bind (nonspecifically) to the DNA molecule, make a transient break in one of the DNA backbones and then pass the other backbone through before resealing and releasing the DNA. This changes the twist, and thus for a circular or topologically constrained molecule, changes the writhe.

One example where type I topoisomerases naturally come into play is when the DNA helix is unzipped, for example during DNA replication (copying). If the molecule is constrained at the end where the unzipping fork is heading, then supercoils build up in advance of the unzipped region. If a type I topoisomerase does not release the supercoils, then eventually the torsional strain becomes too extreme and the DNA molecule breaks.

Type I topoisomerases are found in all organisms studied thus far, and loss of these proteins is lethal to the cell.

---

[2]We will consider proteins, imprecisely, as small molecular machines that act on DNA. Like DNA, they are biopolymers. The fundamental unit is an amino acid (e.g. lysine, tyrosine or glutamine), rather than a nucleotide. There are 20 different flavours of amino acids, and the exact composition and ordering of these amino acids determines the shape and function of the protein.

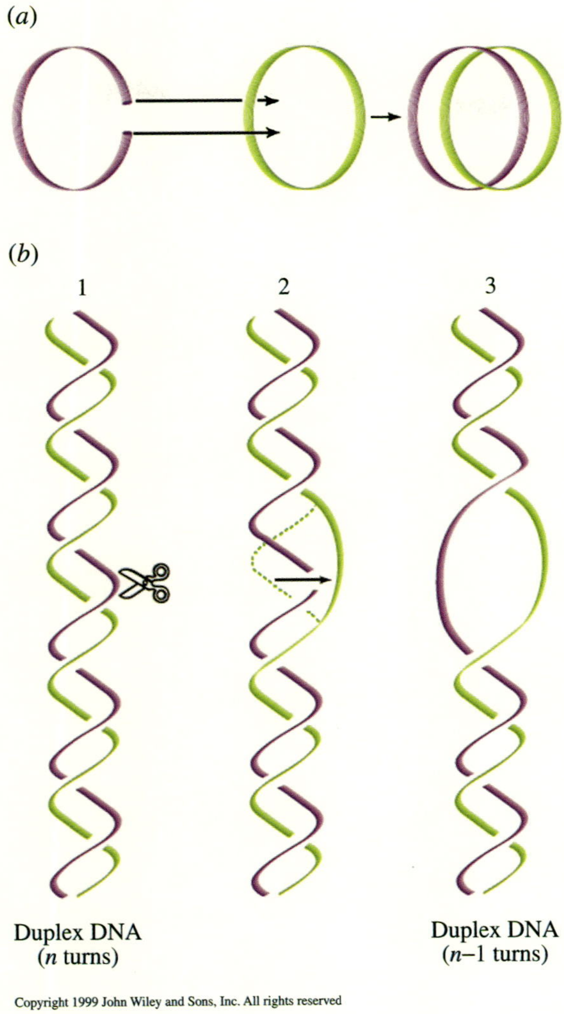

FIGURE 4. Action of Topoisomerase I. From *Biochemistry* by Donald Voet and Judith G. Voet (2004). Reproduced with permission of John Wiley and Sons Ltd.

5.2. Regulating DNA knotting and linking: Type II topoisomerases. Like type I topoisomerases, Type II topoisomerases are also both ubiquitous and essential. Their primary function is to change DNA knot or link type, but they can also remove or (in the case of DNA gyrase) add DNA supercoils.

One example where type II topoisomerases naturally come into play is when circular DNA (such as bacterial genomic DNA) is replicated. The end result of this process is two linked circular molecules. If a type II topoisomerase does not unlink the two daughter molecules, then the bacterial cell cannot divide properly and commits suicide.

Because of their crucial role in many cellular processes, type II topoisomerases have been drug targets for both human cancer and infectious diseases [**38, 61**]. For example, since bacterial type II topoisomerases differ from human, many antibiotics work by inhibiting these proteins, killing the bacterial infection by inhibiting unlinking of daughter DNA as described above. Thus from both a basic biochemical and pharmaceutical perspective, understanding the mechanism by which type II topoisomerases act has been very important.

5.3. Mechanism of type II topoisomerases. One might assume that type II topoisomerases, as small molecules, act independently of the global DNA topology. Surprisingly, however, they act in a manner that preferentially *un*knots and *un*links DNA—this is known in the literatures as topological simplification. Furthermore, type II topoisomerases simplify DNA topology in an extraordinarily efficient manner [**80**]. Exactly how this is achieved has been intensely debated.

What is generally agreed on is that a generic type II topoisomerase binds to helix-helix juxtapositions, such as a supercoil or knot/link crossing (see Figure 5). The first helix, called the G (for Gate) segment is broken to allow the T (Transported) segment to pass in a unidirectional manner [**3**]. The G segment is then resealed, resulting in a crossing change (from $+1$ to -1 or *vice versa*). This process relies on ATP hydrolysis. [**3**]. A beautiful recent crystal structure of a type II topoisomerase bound to the G segment elegantly demonstrates that a type II topoisomerase severely bends the DNA (approximately 150°) [**29**]. This supports models that incorporate DNA bending [**11, 99**]. We discuss two in some detail below.

The first model, jointly proposed by a topologist (G. Buck) biologist (L. Zechiedrich) [**11**], describes how the local information obtained at a crossing can be used to determine the global topology. They classify crossings in terms of several vector parameters, and show that certain crossing types, 'hooked junctions', are associated to nonsupercoiled crossings. Crossing changes at these hooked junctions are then more likely to lead to topological simplification.

The second model [**99**] by Vologodskii and co-workers, gives numerical (Monte Carlo simulations) and indirect experimental evidence that a type II topoisomerase actually bends the G segment into a hairpin conformation upon binding. There is some experimental evidence that DNA knots localize, which would lead to the T-segment inside the hairpin more often for knots than for unknots.

5.4. Topological and numerical strategies to understand type II topoisomerase unknotting. Topologically, understanding type II topoisomerase unknotting is part of a larger movement to understand what knots can be obtained from a given one via a single crossing change, that is, by a single round of topoisomerase relaxation. In general, this is very large question—for example, one area of knot theory research has been to (so far incompletely) classify all knots with 'unknotting number 1'—that is, all knots that become the unknot after a single crossing change.

More generally, the Gordian or strand passage distance between any 2 links or knots K and L, $d(K, L)$ is defined to be the minimum number of crossing changes ($+1$ to -1 or *v.v.*) needed to convert K into L, taken over all projections. It is a metric on the space of knot types—e.g. $d(K, L)$ is less than or equal to the sum of their unknotting numbers.

In this framework, we can consider which knots L are adjacent to a given knot K—i.e. L, s.t. $d(K, L) = 1$ or equivalently which knots L are a single TopoII move from K. Recently, there has been a flurry of activity in this area of 'knot

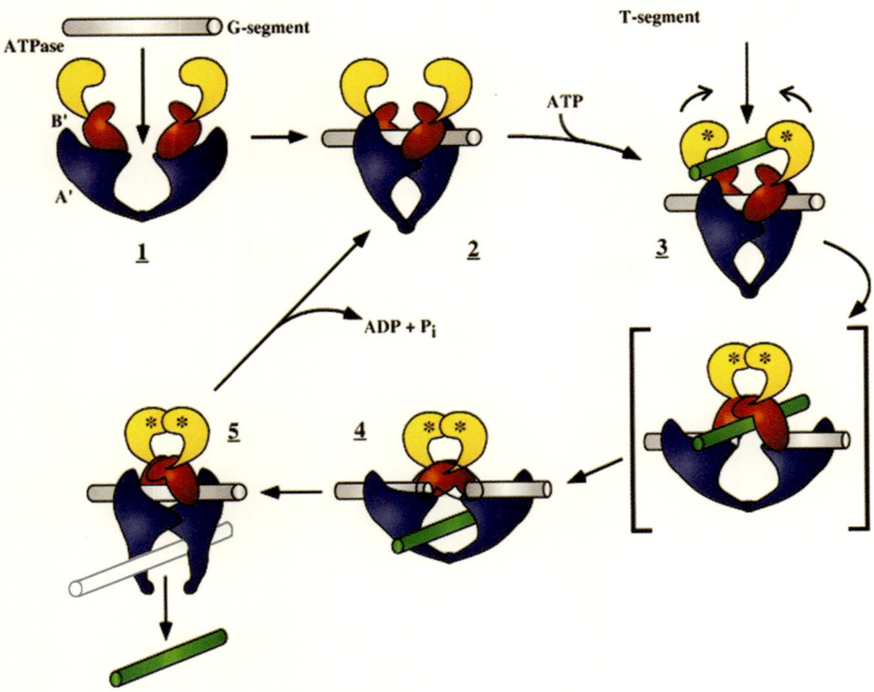

FIGURE 5. Rough mechanism of a type II topoisomerase. Reprinted by permission from Macmillan Publishers Ltd.: *Nature*, Structure and mechanism of DNA topoisomerase II. J. M. Berger, S. J. Gamblin, S. C. Harrison, J. C. Wang (1996).

adjacency'. The problem has been attacked using both finite-type invariants and Dehn surgery techniques. There has been some progress using both, in particular to determine bounds on adjacency number in terms of knot genus. (Recall that the genus g of a knot K is the minimal genus of an orientable spanning surface for K.) One example of this is via knot concordance, since distance g knots can be thought of as together bounding a smoothly embedded genus-g surface in 4-space. Applications to unknotting number and knot adjacency arise from the Heegaard-Floer knot homology of Ozsvath-Szabo and Rasmussen and from Khovanov-type knot homologies (due to work of Rasmussen [**76**] and Lobb [**64**]). Each of these theories are now combinatorial in nature, and so provide computable lower-bounds for knot adjacency. In particular, they have been used to classify all knots K with 9 or fewer crossings and $u(K) = 1$ [**71**].

Because DNA is plectonemically supercoiled, then some of the most common DNA knots and links are members of the *4-plats*: knots or 2-component links that admit a projection consisting of a braid on 4 strings, with one strand free of crossings[3]. Considering the smaller question for 4-plat knots and links has been more tractable, and there has been quite a bit of progress in this area already. For example, Kanenobu and Murakami classified all unknotting number one 4-plat knots [**52**], and Kohn classified all unlinking number one (necessarily 2-component) 4-plat

[3]We will discuss 4-plats in more detail in Section 8 below.

links [**55**]. More generally, a classification of adjacent 4-plats was given independently by Torisu, Darcy and Sumners, and John Berge [**93, 24**]. So in theory one could determine exactly when two knots were related by a single TopoII move, and the distance between any two 4-plats (i.e. the minimum number of times TopoII must act to convert K into a different given 4-plat type). For approximately 20 low-crossing knots and their mirror images, Darcy has a table determining these distances [**24**]. On the Heegaard-Floer homology side, although *a priori* the homologies are harder to compute the higher the crossing number of K, for 2-bridge knots the computations are easier (for example [**77**]).

In parallel, numerical simluations by Stasiak and collaborators have investigated the probability of K being converted into L (where possibly $K \simeq L$) via a single type II topoisomerase [**89**]. Loosely, they begin with a closed knotted polymer chain, randomize it via crookshank moves, and then allow 1 more move that could (or could not) involve a single strand passage. They then calculate the new knot type. Some interesting things emerge—*e.g.*, in 95% of the cases for the unknot, a single intersegmental passage returned the unknot again.

They (nicely) imagine the knot space as a foam, with a crossing change corresponding to moving between two adjacent bubbles. Then they consider the areas between different neighbouring knot spaces, in particular, whether there is a relation between the physical 3-D space available for configurations of a given knot and the high-dimensional configuration space occupied by the same knot type. They argue that only the probability to maintain the knot topology (*i.e.* the unknot remaining the unknot after a single crossing change) is related to the Length/Diameter ratio of the ideal configuration of the knot. This probability gave them a way to estimate the surface-to-volume ratio of configuration spaces of various knots.

6. Determining DNA topology

As discussed in Section 2.2, DNA knots and links (*aka* links) have been implicated in a number of cellular processes (see [**65, 18**] and references therein). They can occur as a result of replication (discussed above) and recombination (discussed below). Knots and links also arise as the *products* of, i.e. as the result of enzyme action. In addition to the topoisomerases, discussed above, knots and links also arise as products of reactions from two other protein families, the recombinases and transposases [**18, 72**]. Most prevalently, DNA knots and links arise as products of certain laboratory experiments (called 'topological enzymology' experiments) on artificially constructed small (5–10 kb) circular molecules [**101, 47, 66, 51, 22, 48, 23, 20, 86, 16**]. These product knots and links can help determine the binding and mechanism of the protein being studied. The variety of DNA knots and links observed has made biologically separating and distinguishing these molecules a critical issue.

Knotted DNA was first discovered in the lab in 1981 by Liu and Davis [**63**]. Experimentally, DNA knots and links can be resolved in two ways: via electron microscopy or electrophoretic migration [**56, 92, 107**].

DNA molecules have been visualized beautifully by electron microscopy. (See Figure 6.) In this process, the entire DNA molecule is ensheathed in a (RecA) coating, which thickens and stiffens the molecule. It is then possible to determine the sign of the axis crossings and thus definitively determine the precise knot or link type. However, this process can be labourious and difficult, particularly producing a relatively large amount of knotted/linked DNA and deciphering the sign of crossings. Perhaps unsurprisingly then, there are only a few dozen such published electron microscopy images of DNA knots and links, and only a few labs (most notably Andrzej Stasiak's) that have developed the necessary expertise.

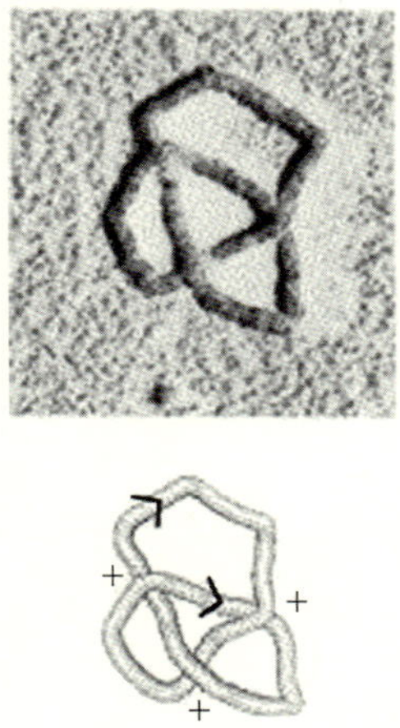

FIGURE 6. Electron microscope image of knotted DNA. Courtesy: Shailja Pathania

A much more widespread (but incomplete) technique to separate DNA knot and link types is that of *agarose gel electrophoresis*. To do this, the DNA is first nicked—that is, one backbone strand is cut to release any supercoils. Next, one prepares a slab of jello-like substance from agarose, a seaweed derivative. This gel slab is then placed in an aqueous bath, a current is run through the bath, and DNA is pipetted into one end of the gel. Recall from Section 2.1 that the DNA backbones are composed of alternating sugar and phosphate groups. These phosphate groups are negatively charged, and so the DNA will migrate to the positive end of the gel. If all the nicked DNA pipetted in is of the same molecular mass and sequence, then this process, *gel electrophoresis*, will stratify DNA knots and links. Under UV light, this appears as a series of dark bands marching down the white gel. Each band corresponds to many DNA molecules, and its intensity is proportional to the amount of DNA present.

Gel electrophoresis is straightforward and requires relatively small amounts of DNA. Typically the distance a given knot or link migrates through the gel is proportional to the minimal crossing number (MCN, the fewest number of crossings with which it can be drawn). Under standard conditions, knots of greater MCN migrate more rapidly than those with lesser MCN [**53, 91, 60**][4]

However, there are 1,701,936 knots with MCN ≤ 16, so a better stratification is needed to positively identify a particular knot [**50**]. By cleverly running the DNA in a second dimension, 2-dimensional gel electrophoresis can separate some prime knots with the same MCN [**92**]. Unfortunately, there is no clear relationship between relative migration of knots with the same MCN in the second dimension.

[4]However, there are gel conditions where, for example, the unknot will migrate ahead of the trefoil.

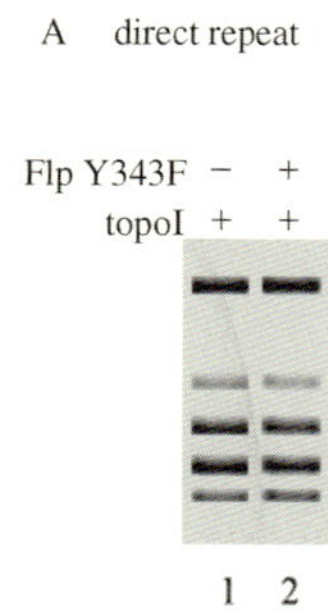

FIGURE 7. Gel electrophoresis of linked DNA. In this image, the DNA was travelling towards the bottom of the page. (Author's Gel.)

In some cases, for DNA of a given length, (1-dimensional) gel electrophoresis can separate some knots with the same MCN. For example, the five and seven-crossing torus knots migrate more slowly than the corresponding five and seven-crossing twist knots [**53, 97**]. This has not generalized, although recent experiments tantalizingly indicate that knots and links may migrate linearly with respect to the average crossing number of a particular conformation—the *ideal configuration*[5] of the knot or link [**97, 58, 54**].

To further complicate matters, gel electrophoresis is incomplete in that adjacent bands determine only relative MCN or average crossing number, not precise values. That is, give two DNA molecules A and B, if A runs further through the gel than B, then A has a higher MCN than B, but one cannot immediately generalize whether MCN(A) = MCN(B) + 1 or MCN(A) = MCN(B) + 2, etc. Nor can one determine if MCN(B) = 3, say, or 4 without further analysis.

Thus gel electrophoresis and electron microscopy are often used in tandem. First DNA is run through a gel to isolate a knot with a given MCN, and the particular band of interest is physically cut out of the gel. This DNA is further purified, and then examined via electron microscopy to determine the precise knot or link type.

To attempt to determine the exact knot or link type without resorting to electron microscopy, one must also construct an appropriate control: a knot ladder. This is a solution of DNA molecules whose precise MCN are known, which is then pipetted into an adjacent well to the unknown DNA in the gel. These known knots or links then serve as markers in the gel in which to calibrate the DNA molecules of interest. While this can be done in some cases (*e.g.* T4 topoisomerase will produce a ladder of twist knots [**100**]), generating such a ladder of known knots and links from DNA of the same length and similar sequence as the unknown knots is highly nontrivial.

Because of the challenges inherent in both electron microscopy and gel electrophoresis, new methods for determining (or predicting) the precise DNA knot or

[5]*Ideal geometric configurations* of knots are the conformations that allow maximal radial expansion of a virtual tube of uniform diameter centered around the knot [**45, 54**]. There is a growing amount of mathematical literature on these ideal configurations. Originally they were determined computationally, but recent analytic work ([**15**] and separately [**41**]) has proven the existence and uniqueness of these solutions.

link type has been an active arena for both mathematicians and experimentalists (see below). More generally, modelling DNA migration through a gel is an active area of research.

7. Changing DNA topology: Site-specific recombination

As discussed above, the sole function of type II topoisomerases is to change DNA knot or link type. We now turn our attention to knots and links that arise from the action of another family of proteins, the *site-specific recombinases.* These proteins mediate *site-specific recombination*, the reshuffling of the genetic sequence, for example changing GATTACA into ACATTAG.

Site-specific recombination is important because of its key role in a wide variety of biological processes. The result of site-specific recombination is the deletion, insertion or inversion of a DNA segment. (See Figure 8.) This corresponds to a wide variety of physiological processes, including crucial steps in viral infections.

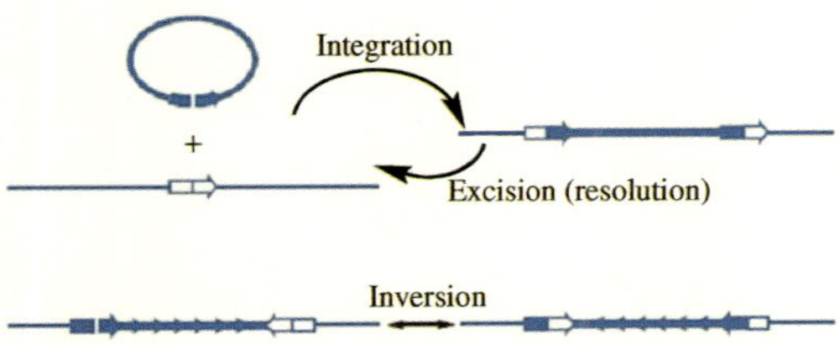

FIGURE 8. Outcomes of site-specific recombination.

In addition to their inherent biochemical interest, pharmaceutical and agricultural industries have become increasingly involved in genetically modifying organisms or testing whether a mutation in a particular gene leads to a disease. As a result, these industries are now interested in site-specific recombinases as tools for precisely manipulating DNA (*e.g.* [**36**]).

While changing DNA topology is not the primary function of these site-specific recombinases, it can be a byproduct of the reaction. If the original circular DNA is supercoiled, the supercoils can be converted into knot or link crossings during the process of recombination. (See [**7**] or [**21**] for more information.)

7.1. The mechanics of site-specific recombination. Minimally, site-specific recombination requires both the site-specific recombinase and two short (30–50bp) DNA segments, the *crossover sites*, inserted into 1 or 2 small circular DNA molecules [**44**]. If there are 2 crossover sites on a single molecule of circular DNA, they can be in either *direct* orientation (head-to-tail, *e.g.*, ...ATGC...ATGC) or *inverted* orientation (head-to-head *e.g.* ...ATGC...CGTA). (See Figure 9.) Larger site-specific recombination systems can also require additional proteins and DNA sites.

During site-specific recombination, two recombinase molecules first bind to each crossover site. The two crossover sites are then brought together within a *recombinase complex*, B: the smallest convex region containing the four bound recombinase molecules and the two crossover sites [**88**]. So B is a *topological ball* (*i.e.*, it can be deformed to a round ball). The crossover sites can be located either on the outside or inside the 4 recombinase subunits [**59, 70, 66, 62, 30, 46, 78, 4**].

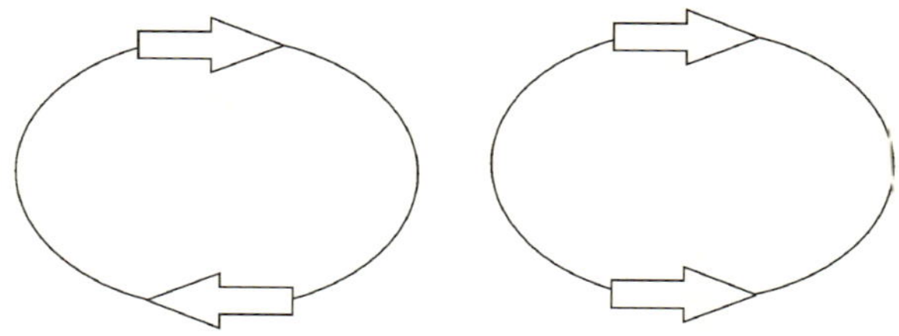

FIGURE 9. On a circular DNA molecule, the two crossover sites can be in direct (left image) or inverted orientation.

Site-specific recombination roughly has three stages. First, two recombinase molecules bind to each of two specific sites on one or two molecules of circular DNA (known as the *substrate*) and then bring them close together. The sites and bound proteins together are called the *recombinase complex*. Next, the sites are cleaved, exchanged and resealed. The precise nature of this intermediary step is determined by which of the two recombinase subfamilies the particular protein belongs to. And finally, the rearranged DNA, the *product*, is released.

Multiple rounds of strand exchange can occur before releasing the DNA—this process is known as *processive* recombination. This is in contrast to *distributive* recombination, where multiple rounds of the entire process of recombination (including releasing and rebinding) occurs.

7.2. The two families of site-specific recombinases. Site-specific recombinases fall into two families—the serine (also known as the resolvase) and tyrosine (also known as the integrase) recombinases—based on the particular amino acid in the protein polymer that catalyzes the cleaving reaction [**44**]. The serine and tyrosine recombinases also differ in their mechanism of cutting and rejoining DNA at the crossover sites. Both families are large: an evolutionary analysis has been performed on 72 serine recombinases [**87**] and a recent sequence search documents approximately 1000 related sequences of putative tyrosine recombinases [**2**].

The diverse family of serine recombinases is comprised of four subfamilies[6] [**87**]. These recombinases may trap a fixed number of supercoils before initiating recombination [**44**]. For example, Tn3 resolvase requires three negative supercoils to be trapped by the binding of (nonactive) resolvase molecules. These trapped supercoils (outside of the recombinase complex) together with the recombinase complex itself are known as the *synaptic complex* [**101, 88**]. Likewise, the invertases also require a fixed number of supercoils trapped outside the recombinase complex. Rather than using additional recombinase molecules, they rely on additional proteins (called accessory proteins) and DNA sites (called enhancer sequences), which facilitate the organization of a unique synaptic complex that promotes DNA cleavage.

With serine recombinases, recombination proceeds through a concerted 4-backbone cleaving and rejoining reaction [**44**]. Serine recombinases can perform processive recombination.

In contrast, tyrosine recombinases first cleave, exchange and reseal two sugar-phosphate backbones. The DNA-protein complex then proceeds through an intermediary structure (a Holliday junction) before repeating the process with the other

[6]resolvases (such as Tn3 and $\gamma\delta$), invertases (such as Gin, Hin, Pin, and Min), large serine recombinases (also called large resolvases) and IS elements.

two DNA backbones [**43, 44**]. Most tyrosine recombinases, including the most famous members Flp, λ Int and Cre, tolerate varying numbers of supercoils outside of the recombinase complex. (However, there are exceptions, most notably XerCD, which trap a fixed number of supercoils using accessory proteins before initiating cleavage [**16**]). Like serine recombinases, tyrosine recombinases can also employ accessory proteins to help assemble the synaptic complex, and to drive the overall reactions (*e.g.* λ Int and XerCD) [**16, 4**].

7.3. DNA knots and links as recombination products. As mentioned above, site-specific recombination can transform supercoiled circular DNA into a knot or link. Given a substrate with crossover sites in a particular orientation, a single round of recombination mediated by a serine recombinase yields a unique topological product. For example, Tn3 acting on an unknot with direct sites yields exclusively the (2,2)-torus link (the Hopf link). Multiple rounds of processive recombination successively lead to the figure eight knot 4_1, the Whitehead link 5_1^2 and the six-crossing knot 6_2. By contrast, a single round of recombination mediated by a generic tyrosine recombinase yields a spectrum of DNA knots or links[7]. For example, Flp acting on an unknot with direct sites yields the unlink and a variety of 2-torus links: $2_1^2, 4_1^2, 6_1^2, 8_1^2, 10_1^2, 12_1^2$, and higher-crossing links as well. This distribution of knot/link products should reflect the supercoiling density of the substrate DNA.

Understanding precisely which knots and links arise during site-specific recombination can help understand the details of the process, *e.g.* [**42**]. Topological techniques have played a significant role in characterizing knotted and linked products of site-specific recombination. For example, several approaches have been developed to determine a particular DNA knot or catenane type, including utilizing the node number for knots [**18**], the Jones polynomial for catenanes [**6**], Schubert's classification of 4-plats [**104**] and the HOMFLY polynomial [**106**]. Below, we discuss two large topological contributions to our understanding of site-specific recombination.

7.4. Predicting DNA knot and link products. Recently, Buck and Flapan [**8, 7**] developed a predictive model for DNA knots and links that arise as products of site-specific recombination. More specifically, rather than focusing on a specific recombinase as many earlier studies have done, the authors presented a topological model that predicts which knots and links can occur as products of site-specific recombination *in general.* They do this by describing the topology of how DNA knots and links are formed as a result of a single—or multiple rounds of processive—recombination event(s), given a plectonemically supercoiled unknot, unlink, or $T(2, m)$ torus knot or catenane substrate. (See Figure 10 for illustrations of each substrate type.)[8]

The general idea is to consider a ball B containing the convex hull of the four recombinase molecules, and a spanning surface D (consisting of a disc for the unknot, two disjoint discs for the unlink, or twisted band for a torus knot or link) whose boundary is the unrecombined DNA axis. Then consider $D \cap B$ pre-recombination and post-recombination. Similarly, characterize $D \cap cl(S^3 \setminus B)$ for each substrate. Restrict these possibilities using topological arguments. Then glue

[7]There are a few exceptions to this, for example the tyrosine recombinase XerCD. However, in this introductory article we will consider only the generic case.

[8]Note: *All* figures represent the axis of duplex DNA.

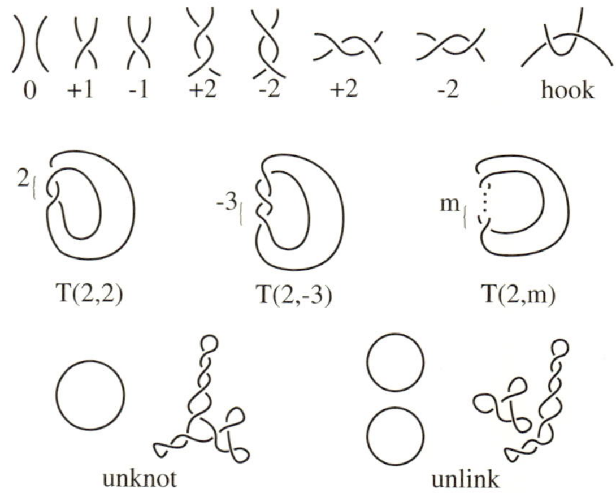

FIGURE 10. *Above:* Possible conformations of the recombinase complex. *Below:* Possible substrates: the unknot, unlink, or torus knot or link.

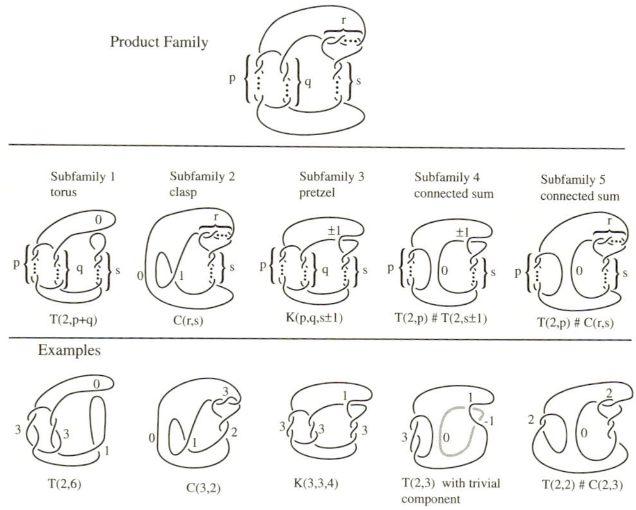

FIGURE 11. Product family, and subfamilies. Note that this family contains the same torus knots and links as subfamily 1.

each of the post-recombinant forms of $D \cap B$ to each form of $D \cap cl(S^3 \setminus B)$ to classify possible product knots and links.

The model is independent of the size of the substrate, the sign of the supercoils and the site orientation. The model relies on three assumptions, each supported by biological evidence. Given these assumptions, the authors predict that products arising from site-specific recombination must be members of a single family of products (illustrated in Figure 11). This is a specific family of Montesinos knots and links, described in more detail in Section 8.1, given by $(\frac{1}{p}, \frac{1}{q}, \frac{r}{rs+1})$, with $p, q, r, s \in \mathbb{Z}$.

One comforting thing about this family is its small size. As mentioned above, the minimal crossing number (MCN) of a DNA knot or catenane can be determined experimentally [**60**]. For small values of the MCN there are not many knots or links with a given value. However, the number of knots and links with MCN $= n$ grows exponentially as a function of n [**33**], and there are 1,701,936 knots with MCN $\leq$ 16 [**50**]. So knowing the MCN is not sufficient to determine the knot or catenane.

However, as proved in [**8**], the total number of knots and links in the family $F(p, q, r, s)$ of Figure 11 grows linearly with n^3. So the proportion of all knots and links which are contained in our family decreases exponentially as n increases. Thus, knowing the MCN of a product and knowing that the product is in one of this families allows us to significantly narrow the possibilities for its knot or link type.

There are several applications of this work. Firstly, it can predict knot and link products for previously uncharacterized data. Secondly, it can determine pathway of recombination: processive (multiple strand exchanges before releasing) versus distributive (entire recombination process repeats). In particular, it predicts that any knot or link product that is not of the family above must arise distributively. And finally, this work can help determine sequence of products of processive recombination.

This recent work complements earlier results of [**90**], which used the tangle model described below and several biologically reasonable assumptions to solve tangle equations. They then determined which (4-plat) knots and links arise as a result of (possibly processive) site-specific recombination on the unknot for the serine subfamily of recombinases. In addition to an unknotted substrate for a generic recombinase, the more recent work allows substrates that are unlinks with one site on each component, as well as $(2, m)$-torus knots and links. The more recent work of [**8, 7**] also does not assume the tangle model holds or that the products must be 4-plats. This is particularly important as (distributive) recombination has been seen to produce knots and links which are connected sums, and thus not 4-plats.

8. The tangle model

We now consider the alternative scenario. Suppose we know the substrate and product topologies for a given site-specific recombinase. For example, as above, we have determined that Flp acting on an unknot with direct sites yields the unlink and a variety of 2-torus links: $2_1^2, 4_1^2, 6_1^2, 8_1^2, 10_1^2, 12_1^2$. Our goal is now to use this input, together with topological arguments to determine details of the pathway and/or mechanism. For example in [**42**], using the topological products as probes, it was determined that Flp aligns the crossover sites so that their sequences are aligned antiparallel to each other.

In the tangle model different regions of the DNA molecule are represented by tangles, and the action of the recombinase as a change in one of the tangles. Given the substrate and product topologies, mathematicians can help biologists by finding all tangle combinations that may explain the recombinase's action. Some of these mathematically possible solutions can then be eliminated through biological considerations. Here, the input is the knot or link type of both the substrate and product DNA. The output of the tangle model is a list of solutions to the corresponding tangle equations, which yield a corresponding set of solutions for the recombination pathway and/or mechanism.

The tangle model was originally developed by Ernst and Sumners to describe the action of particular site-specific recombinases in terms of tangle sums [**34**]. Building on the experimental work of Wasserman and Cozzarelli [**101**] and Conway's theory of tangles [**19**], they used their tangle model to make predictions—later experimentally verified—about how the recombinase Tn3 resolvase interacts with DNA [**34**]. The tangle model has since been used to determine various features of protein-DNA interactions for a number of specific proteins [**9, 10, 23, 24, 96, 42, 35, 90, 95, 94**].

8.1. Tangles, tangle operations and double branch covers. We begin by stating a few elementary facts about tangles. (For a more comprehensive introduction to tangles, see [**9**] and [**40**].) A *tangle* T is (the isotopy (rel ∂) class of) a pair (B^3, t), where B^3 is a 3-ball with a given boundary parametrization with four distinguished boundary points labelled NW, NE, SW, SE, and t consists of a pair of properly embedded unoriented arcs with endpoints NW, NE, SW and SE. We say two tangles C and D are *equivalent* if there exists an isotopy taking C to D, which remains the identity on ∂C.

Tangles can be divided into three mutually exclusive families: locally knotted, rational and prime. See Figure 12.

FIGURE 12. From left to right, a rational, prime and locally knotted tangle.

A tangle is *locally knotted* if there exists a sphere in B^3 meeting t transversely in 2 points such that the 2-ball bounded by the sphere intersects t in a knotted spanning arc.

Rational tangles are the second family; they are so-called because their isotopy classes are in one-to-one correspondence with the extended rational numbers ($\mathbb{Q} \cup \{\infty\}$) via a continued fraction expansion, as first constructed by Conway [**19**]. (See [**12**] for an outline of the original proof, based on branched covering spaces. Alternatively see [**40**] for subsequent nice classifications that utilize less machinery.) A tangle whose corresponding rational number is $\frac{p}{q}$ will be denoted by $\left(\frac{p}{q}\right)$. Rational tangles are formed by an alternating series of horizontal and vertical half-twists of two (initially untwisted) parallel arcs (and hence are freely isotopic to them). Any continued fraction decomposition of $\frac{p}{q} = a_n + 1/(a_{n-1} + ... (1/a_1))$ yields a finite list of integers $[a_1, \ldots, a_n]$ which tell us how to twist the strands around each other to get a diagram of the tangle. The (0) tangle corresponds to two untwisted horizontal arcs (one joining NE to NW and one joining SE to SW), whereas the (∞) tangle corresponds to two untwisted vertical arcs. The double branch cover of a rational tangle is a solid torus.

All locally unknotted, nonrational tangles are *prime*. Bleiler proved that the minimal prime tangle has a minimal projection with five crossings [**5**].

There are several operations one can perform on tangles. We concentrate on two. (See Figure 13.)

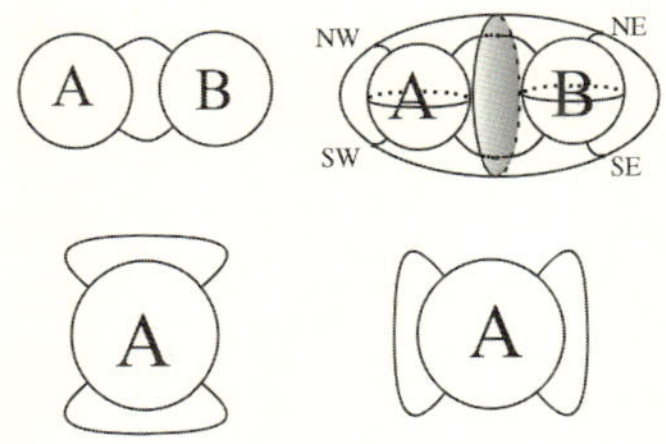

FIGURE 13. Two tangle operations: sum and numerator closure.

The first operation forms a knot or 2-component link from a given tangle A: the *numerator closure*, $N(A)$. This adds an unknotted arc joining the northern endpoints, and another unknotted arc joining the southern endpoints, or equivalently, the boundary of A and (0) are identified so that E_A is identified with $\mathrm{E}_{(0)}$ for $E \in \{NE, NW, SE, SW\}$.

The numerator closure of a rational tangle yields a *4-plat* or *rational knot*: a knot or 2-component link that admits a projection consisting of a braid on 4 strings, with one strand free of crossings [**12**]. Given two rational tangles $\left(\frac{p}{q}\right)$ and $\left(\frac{p'}{q'}\right)$, then $N\left(\frac{p}{q}\right) = N\left(\frac{p'}{q'}\right)$ iff $p = p'$ and $q^{\pm 1} \equiv q \mod p$. Thus given a 4-plat, we can write it as the numerator closure of a rational tangle that is unique up to the relationship above. Schubert showed that all 4-plats are prime knots [**81**]. We will denote the 4-plat obtained by the numerator closure of $\left(\frac{p}{q}\right)$ as $b(p,q)$. For example, we can write the unknot as $b(1,1)$, and the trefoil as $b(3,1)$.

The second operation, *tangle sum*, takes a pair of tangles A, B, and, under certain restrictions, yields a third tangle, $A + B$, by identifying the eastern hemispheric boundary disk of A with the western one of B in such a way that NE_A is identified with NW_B and SE_A is identified with SW_B. Note that the (0) tangle is the identity under this operation: $A + (0) = A$. Beware that under tangle addition, we cannot distinguish between $A + (p)$ added to $(-p) + B$ and A added to B. Thus although tangle summands are written in their simplest form, they are unique only up to an arbitrary number of compensating positive and negative horizontal twists. For example, given a solution $P = (p), R = (r)$ and $O^k = (s)$, then $P = (p+n), R = (r+n)$ and $O^k = (s-n)$ is another solution which is not a minimal projection.

A particular class of prime tangles is obtained by tangle sum of rational tangles *Montesinos* tangles [**68, 67, 69**]. We will use the notation $\left(\frac{a_1}{b_1}, \frac{a_2}{b_2}, \ldots, \frac{a_n}{b_n}\right)$ to denote the Montesinos tangle obtained by the tangle sum of the rational (possibly integral) tangles $\left(\frac{a_1}{b_1}\right), \left(\frac{a_2}{b_2}\right), \ldots, \left(\frac{a_n}{b_n}\right)$.

Since the sum of a rational tangle and an integral tangle yields a rational tangle [**14**], a Montesinos (nonrational) tangle must have at least two nonintegral summands. The numerator closure of a Montesinos tangle is called a *Montesinos knot or link*. The model discussed in Section 7.4 predicts that all knots and links arising from site-specific recombination fall within a small subfamily of Montesinos knots and links.

Note that $N(A_1 + A_2 + \cdots + A_n)$ is isotopic to the numerator closure of the sum of the same tangles re-ordered by a cyclic permutation and/or a reversal of order. (See Chapter 12 in [**12**] for a primer on Montesinos tangles and knots).

8.2. Double branch covers. Given a tangle T, then we will use $\widetilde{T}$ to mean the double cover of B^3, branched over t. In general, we will write $\text{dbc}(K)$ to denote the three-manifold that is the double cover of S^3 branched over the set K. We now turn our attention to the (compact, connected and orientable) three-manifolds that arise as double branch covers of tangles or 4-plats.

If P is a rational tangle, then $\widetilde{P}$ is a solid torus, which we will denote by V_P. (See Figure 14.)

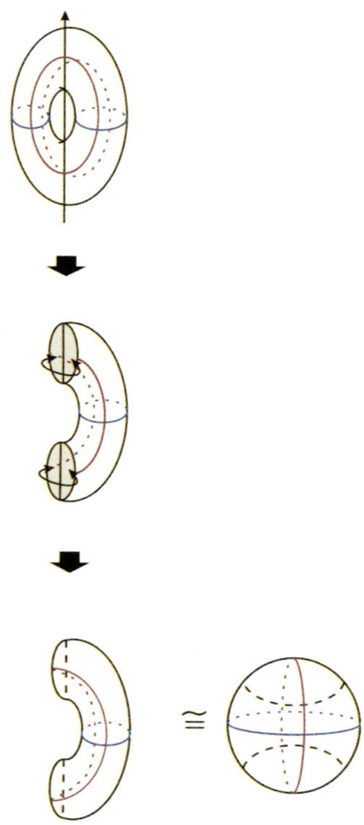

FIGURE 14. Double branch cover of a 3-ball branched over a rational tangle is a solid torus.

Schubert showed that $\text{dbc}(b(p,q))$ is a special type of well-understood three-manifold, the lens space $L(p,q)$. Two 4-plats $b(p,q)$ and $b(p',q')$ are equivalent if and only if their corresponding double branch covers, the lens spaces $L(p,q)$ and $L(p',q')$, are homeomorphic [**81**], so $b(p,q) = b(p',q')$ if and only if $p = p'$ and $q^{\pm 1} \equiv \pm q' \mod p$. (See Rolfsen [**79**], for a good introduction to lens spaces.)

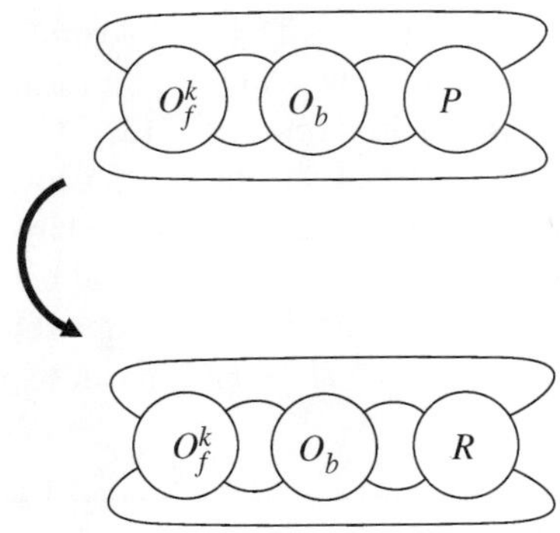

FIGURE 15. Tangle model of recombination: Substrate DNA is the numerator closure of the 3 tangles O_f^k, O_c and P. Recombination is modelled as replacing tangle P with tangle R.

8.3. An illustration of the tangle model.

8.3.1. *Tangle model for a tyrosine recombinase.* We can now describe in full detail a generalization, introduced in [**90**], of the original tangle model of Ernst and Sumners [**34**]. We illustrate this model with a generic member of the tyrosine family of recombinases, the protein *Flp* (pronounced 'flip'). Flp has served as the paradigm for site-specific recombination, and there are a number of proteins (including Cre, and λ Int acting on LR sites) whose products are, topologically speaking, identical to those of Flp.

We model each of the substrates and products as the numerator closure of the sum of three tangles. Each tangle arc represents a segment of double-stranded DNA. In the tangle model pioneered by Ernst and Sumners [**34**], the cutting and joining of DNA is assumed to be completely localized: two of the tangles are unchanged by the action of the protein. In the substrate, the first tangle, P (Parental), represents the two short crossover sites that Flp recognizes and to which it chemically binds and then cuts, rearranges and re-seals. This action can be thought of as removing P and replacing it with a new tangle, R (Recombinant), in the product. The second tangle, O_c, represents the part of the DNA that is physically constrained, but unchanged, by the protein (O stands for Outside and c for constrained). The last tangle, O_f^k, represents the part of the DNA that is free (hence the subscript f) from protein binding constraints. O_f^k can vary depending on the amount of DNA supercoiling present at the time Flp acts. The superscript k indexes these different possibilities. (See Figure 15.)

In terms of tangles, this implies that the substrate and products can be modelled as:

$$N(O_f^k + O_c + P) = \text{substrate (before recombination)},$$

$$N(O_f^k + O_c + R) = \text{product (after recombination)}$$

where $k \in \{0, 1, 2, 3\}$ represents the degree of supercoiling in the substrate. O_f^k varies as k varies, so we obtain different products, as described below.

When Flp acts on DNA it yields a variety of torus knots (for inverted repeats) or links (for direct repeats), depending on O_f^k.

More precisely, when Flp acts on a DNA molecule with inverted sites, experiments have shown that the resulting DNA can be an unknot (with a different DNA sequence), or a knot with up to 11 crossings [**43**]. Crisona *et al.* have obtained images (using electron microscopy) of the simplest products, and have shown that they are the torus knots $b(1,1)$ (the unknot), primarily positive $b(3,1)$ and exclusively positive $b(5,1)$ [**23**]. This experimental evidence indicates that Flp begins with an unknotted DNA substrate with inverted repeats, $b(1,1)$ and converts it via tangle surgery into a torus knot $b(\pm(2k+1),1)$, where $k \in \{0,1,2,3\}$. See Figure 16.

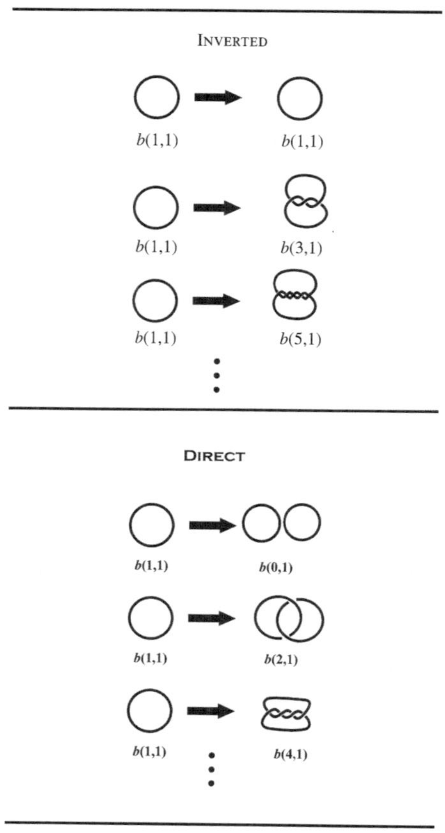

FIGURE 16. Substrate and products of Flp recombination.

We thus model the action of Flp on DNA with inverted repeats as:

Before: $N(O_f^k + O_c + P) = b(1,1) =$ unknot, for $k \in \{0,1,2,3\}$

After: $N(O_f^0 + O_c + R) = b(1,1) =$ unknot,

$N(O_f^k + O_c + R) = b(\pm(2k+1),1) =$ torus knot for $k \in \{0,1,2,3\}$.

8.3.2. *Tangle model for a serine recombinase.* Recall that, in contrast to a generic tyrosine recombinase, proteins in the serine family of recombinases, such as Tn3, require a fixed number of supercoils before they begin cutting and rejoining DNA. Once this requirement is met, they rearrange the DNA, occasionally multiple times, before releasing it. The corresponding tangle equations: from substrate $N(O_f + O_c + P) = K_0$ to products $N(O_f + O_c + nR) = K_n$ are:

$$\begin{array}{ll} N(O_f + O_c + P) = K_0 & \text{for the substrate,} \\ \left.\begin{array}{l} N(O_f + O_c + R) = K_1 \\ N(O_f + O_c + R + R) = K_2 \\ \quad\vdots \\ N(O_f + O_c + \underbrace{R + \cdots + R}_{n}) = K_n \end{array}\right\} & \text{for the products.} \end{array}$$

These equations were first solved (*i.e.*, all constituent tangles have been characterized, given the substrate and product 4-plats) by Ernst and Sumners [**34**]. In particular, they found that $O = O_f + O_c = (-3, 0)$, $R = (1)$ and if P is rational, then $P = (2n, 3, 0)$, for $n \in \mathbb{N}$.

Note, in particular, that since O is always the vertical tangle $(-3, 0)$, then the free part of DNA, O_f, does not vary. This single, fixed O_f is what made the serine recombinase tangle equations more tractable than the generic integrases, whose equations involve a family of tangles O_f^k, indexed by k.

8.4. Strategy. Given the set of tangle equations above, whose products (4-plats) are known, the goal is to determine the constituent tangles (up to equivalence as discussed above). The interplay of tangles and 4-plats with their corresponding double branch covers is the key to many of the results in tangle calculus. For instance, if C and D are tangles, and D is a rational tangle, then the dbc$(C + D)$ is obtained by gluing $\widetilde{C}$ and $\widetilde{D} = V_D$ along annuli that are the lifts of their corresponding gluing disks. If D is integral, then the gluing annulus is boundary reducible, and dbc$(C + D) \simeq \widetilde{C}$ [**9**].

The sum and subsequent numerator closure of two tangles C and D induces a gluing of the boundaries of their respective double branch covers $\widetilde{C}$ and $\widetilde{D}$. If $N(C+D)$ yields a 4-plat $b(p, q)$, then $\widetilde{C} \cup_h \widetilde{D}$ must be the lens space $L(p, q)$, where h is the map that takes $\mu_{\partial\widetilde{C}}$ to $p\lambda_{\partial\widetilde{D}} + q\mu_{\partial\widetilde{D}}$. In particular, when C and D are both rational, $\widetilde{C} = V_C$ and $\widetilde{D} = V_D$ are solid tori, and they form a Heegaard splitting $V_C \cup_h V_D$ of $L(p, q)$.

Replacing tangle P in $N(O + P)$ by tangle R to obtain $N(O + R)$ is called *tangle surgery*. If P and R are rational tangles, then tangle surgery corresponds to replacing V_P with V_R in the double branch cover, and thus corresponds to different Dehn fillings of $\widetilde{O^k}$. In the special case of $N(O^k + P) = b(1, 1)$ (the unknot) where dbc$(b(1, 1)) = S^3$, the tangle surgery corresponds to Dehn surgery on the knot complement $(\widetilde{O^k})$ in S^3. Note by the preceding section the knot is trivial iff O^k is rational.

Ernst and Sumners proved that, given the above equations for a generic integrase, P and R are rational for both direct and inverted repeats [**34**]. More recently in [**9**], Buck and Verjovsky give a different proof for the rationality of R for direct and inverted repeats. So the tangle surgery of replacing P with R corresponds to Dehn surgery in the double branch covers.

Thus the strategy is to use restrictions on the type of Dehn surgeries of $S^3 = \text{dbc}(b(1,1))$ that yield lens spaces. This in turn restricts the possible tangle surgeries, which in turn restricts the exact tangle solutions.

In [**9, 10**], Buck and Verjovsky consider the tangle model for a generic member of the tyrosine recombinase family without any *a priori* assumptions on the constituent tangles. They gave a complete classification of all possible solutions for O_f^k, O_c, P and R to the systems of equations arising from both the direct and inverted cases.

The proofs utilize a fair amount of three-manifold machinery. In particular, they utilize the cyclic surgery theorem, recent work of Kronheimer, Mrowka, Ozsváth and Szabó [**57**], and work of Ernst [**31, 32**].

The results in [**9, 10**] helped illuminate some of the biochemical steps of site-specific recombination. In particular, the elucidation of the tangle O_c helped determine that Flp aligned the sites in an antiparallel alignment [**42**].

9. Conclusion and future directions

There has been dramatic progress in understanding site-specific recombination through topological probing. For a generic tyrosine recombinase, the tangle solutions are completely classified [**9, 10**]. For a number of serine recombinases, the tangle solutions are also known, *e.g.* [**34**].

While the tangle model has been successful in illuminating some of the steps involved, there are many site-specific recombinases whose precise product knot or link type is still uncharacterized and so the tangle model cannot be harnessed. A powerful new strategy combines a predictive model, such as in [**8, 7**], together with the tangle model to probe the recombinase mechanism further.

New avenues in related directions include looking at 3-string tangles, for proteins that use additional sequences in addition to the crossover sites [**26**], and developing a 3-dimensional version of the tangle model [**94**].

By now, the reader is hopefully convinced that DNA Topology has deep biological consequences, *e.g.* for replication and recombination. In particular, DNA knots and links can be dangerous (*e.g.* linked daughter chromosomes) or helpful (*e.g.* as probes for understanding recombination).

We have discussed that it is not (yet) easily possible to understand fully DNA topology via purely biological techniques. For example, electrophoretic migration does not yield the precise knot type, and electron microscopy becomes increasingly difficult as the number of crossings increases. Additionally, the mechanism of type II topoisomerases, whose primary function is to change DNA knot or link type, is still not precisely understood. Thus, mathematicians can help predict DNA topology, for example by restricting the possibilities for recombination products as described in Section 7.4. This in turn can illuminate the biological processes involved.

There are a variety of future research trajectories for both mathematicians and biologists at this rich interface. In addition to the trajectories mentioned earlier, we note that we are only beginning to understand how other DNA-rearranging proteins, such as transposases, form DNA knots and links. Hopefully this article has stimulated the reader to explore—and contribute to!—some of these areas.

References

[1] C. Adams, *Introduction to knot theory*, Applications of Knot Theory: Proceedings of Symposia in Applied Mathematics (2008).

[2] S.F. Altschul, T.L. Madden, A.A. Schaffer, J. Zhang, Z. Zhang, W. Miller, and D.J. Lipman, *Gapped blast and psi-blast: a new generation of protein database search programs.*, Nucleic Acids Res **25** (1997), 3389–3402.

[3] J.M. Berger, *Type II topoisomerases*, Curr. Opin. Struct. Biol. **8** (1999), 26–32.

[4] T. Biswas, H. Aihara, M. Radman-Livaja, D. Filman, A. Landy, and T. Ellenberger, *A structural basis for allosteric control of DNA recombination by λ integrase*, Nature **435** (2005), 1059–1066.

[5] S. Bleiler, *Knots prime on many strings*, Trans. Amer. Math. Soc. **282** (1984), 385–401.

[6] T.C. Boles, J.H. White, and N.R. Cozzarelli, *Structure of plectonemically supercoiled DNA*, J. Mol. Biol. **213** (1990), no. 4, 931–51.

[7] D. Buck and E. Flapan, *A topological characterization of knots and links arising from site-specific recombination*, J. Mol. Biol. **374** (2007), 1186–1199.

[8] ———, *A topological characterization of knots and links arising from site-specific recombination*, J. Phys. A **40** (2007), 12377–12395.

[9] D. Buck and C. Verjovsky Marcotte, *Tangle solutions for a family of DNA-rearranging proteins*, Math. Proc. Cambridge Philos. Soc. **139** (2005), no. 1, 59–80.

[10] ———, *Classification of tangle solutions for integrases, a protein family that changes DNA topology*, Journal of Knot Theory and its Ramifications **16** (2007).

[11] G. Buck and E.L. Zechiedrich, *DNA disentangling by type-2 topoisomerases*, J. Mol. Biol. **340** (2004).

[12] G. Burde and H. Zieschang, *Knots*, De Gruyter Studies in Mathematics, Berlin, 2003.

[13] G. Călugăreanu, *Sur les classes d'isotopie des noeuds tridimensionnels et leurs invariants*, Czech Math. J. **11** (1961), 588–625.

[14] Quach thi Câm Vân, *On a theorem on partially summing tangles by Lickorish*, Math. Proc. Cambridge Philos. Soc. **93** (1983), no. 1, 63–66.

[15] Jason Cantarella, Robert B. Kusner, and John M. Sullivan, *On the minimum ropelength of knots and links*, Invent. Math. **150** (2002), no. 2, 257–286. MR1933586 (2003h:58014)

[16] S.D. Colloms, J. Bath, and D.J. Sherratt, *Topological selectivity in Xer site-specific recombination*, Cell **88** (1997), no. 6, 855–64.

[17] N.R. Cozzarelli, T. Christian-Boles, and James H. White, *Primer on the topology and geometry of DNA supercoiling*, DNA Topology and its Biological Effects (1990), 139–184.

[18] N.R. Cozzarelli, M.A. Krasnow, S.P. Gerrard, and J.H. White, *A topological treatment of recombination and topoisomerases*, Cold Spring Harb. Symp. Quant. Biol. **49** (1984), 383–400.

[19] J.H. Conway, *An enumeration of knots and links, and some of their algebraic properties*, Computational Problems in Abstract Algebra (Proc. Conf., Oxford, 1967), Pergamon, Oxford, 1970, pp. 329–358.

[20] M.M. Cox, *Mobile DNA*, Berg, D.E. and M.M. Howe (eds), ch. DNA inversion in the 2 μm plasmid of Saccharomyces cerevisiae, pp. 661–670, Am. Soc. Microbiology, Washington DC, 1989.

[21] N. Craig, R. Craigie, M. Gellert, and A. Lambowitz (eds.), *Mobile DNA*, ASM Press, Washington, DC, 2002.

[22] N.J. Crisona, R. Kanaar, T.N. Gonzalez, E.L. Zechiedrich, A. Klippel, and N.R. Cozzarelli, *Processive recombination by wild-type Gin and an enhancer-independent mutant. Insight into the mechanisms of recombination selectivity and strand exchange*, J. Mol. Biol. **243** (1994), no. 3, 437–57.

[23] N.J. Crisona, R.L. Weinberg, B.J. Peter, D.W. Sumners, and N.R. Cozzarelli, *The topological mechanism of phage lambda integrase*, J. Mol. Biol. **289** (1999), no. 4, 747–775.

[24] Isabel K. Darcy, *Biological distances on DNA knots and links: applications to XER recombination*, J. Knot Theory Ramifications **10** (2001), no. 2, 269–294, Knots in Hellas '98, Vol. 2 (Delphi).

[25] M.R. Dennis and J.H. Hannay, *Geometry of Călugăreanu's theorem*, Proc. Roy. Soc. A **461** (2005), 3245–3254.

[26] I.K. Darcy, J. Luecke, and M. Vazquez, *Tangle analysis of difference topology experiments: applications to a Mu protein-DNA complex*, arXiv:0710.4150v1 (2007).

[27] R.E. Depew and J.C. Wang, *Conformal fluctuations of the DNA helix*, Proc. Natl. Acad. Sci. USA **72** (1975), no. 11, 4275–4279.

[28] De Witt L. Sumners (ed.), *New scientific applications of geometry and topology*, American Mathematical Society, Providence, RI, 1992, Papers from the American Mathematical Society Short Course held in Baltimore, Maryland, January 6–7, 1992.

[29] K.C. Dong and J.M. Berger, *Structural basis for gate-DNA recognition and bending by type IIA topoisomerases*, Nature **450** (2007), 1201–1206.

[30] E. Ennifar, J.E.W. Meyer, F. Buchholz, A.F. Stewart, and D. Suck, *Crystal structure of a wild-type Cre recombinase-loxP synapse reveals a novel spacer conformation suggesting an alternative mechanism for DNA cleavage activation*, Nucleic Acids Res. **31** (2003), no. 18, 5449–5460.

[31] C. Ernst, *Tangle equations*, J. Knot Theory Ramifications **5** (1996), no. 2, 145–159.

[32] ———, *Tangle equations. II*, J. Knot Theory Ramifications **6** (1997), no. 1, 1–11.

[33] C. Ernst and D. W. Sumners, *The growth of the number of prime knots*, Math. Proc. Cambridge Philos. Soc. **102** (1987), no. 2, 303–315.

[34] ———, *A calculus for rational tangles: applications to DNA recombination*, Math. Proc. Cambridge Philos. Soc. **108** (1990), no. 3, 489–515.

[35] ———, *Solving tangle equations arising in a DNA recombination model*, Math. Proc. Cambridge Philos. Soc. **126** (1999), no. 1, 23–36.

[36] R. Feil, *Conditional somatic mutagenesis in the mouse using site-specific recombinases*, Handbk. Exp. Pharmacol. **178** (2007), 3–28.

[37] M.D. Frank-Kamenetskii, *Unraveling DNA*, VCH Publishers, New York, NY, 1993.

[38] S.J. Froelich-Ammon and N. Osheroff, *Topoisomerases poisons: harnessing the dark side of enzyme mechanism*, J. Biol. Chem. **270** (1990), 21429–21432.

[39] F.B. Fuller, *The writhing number of a space curve*, Proc. Natl. Acad. Sci. USA **68** (1971), 815–819.

[40] J.R. Goldman and L. H. Kauffman, *Rational tangles*, Adv. Appl. Math. **18** (1997), 300–332.

[41] O. Gonzalez and R. de la Llave, *Existence of ideal knots*, J. Knot Theory Ramifications **12** (2003), no. 1, 123–133. MR1953628 (2003j:57010)

[42] I. Grainge, D. Buck, and M. Jayaram, *Geometry of site alignment during Int family recombination: Antiparallel synapsis by the Flp recombinase*, J. Mol. Bio. **298** (2000), 749–764.

[43] I. Grainge and M. Jayaram, *The integrase family of recombinase: organization and function of the active site*, Mol. Microbiol. **33** (1999), no. 3, 449–56.

[44] N. D. Grindley, K. L. Whiteson, and P. A. Rice, *Mechanisms of site-specific recombination*, Ann. Rev. Biochem. **75** (2006), 567–605.

[45] A. Y. Grosberg, A. Feigel, and Y. Rabin, *Flory-type theory of a knotted ring polymer*, Phys. Rev. E Sta.t Phys. Plasmas Fluids Relat. Interdiscip. Topics **54** (1996), no. 6, 6618–6622.

[46] F. Guo, D.N. Gopaul, and G.D. van Duyne, *Structure of Cre recombinase complexed with DNA in a site-specific recombination synapse*, Nature **389** (1997), no. 40–46.

[47] K.A. Heichman, I.P. Moskowitz, and R.C. Johnson, *Configuration of DNA strands and mechanism of strand exchange in the Hin invertasome as revealed by analysis of recombinant knots*, Genes Dev **5** (1991), no. 9, 1622–1634.

[48] R. H. Hoess, A. Wierzbicki, and K. Abremski, *The role of the loxP spacer region in P1 site-specific recombination*, Nucleic Acids Res. **14** (1986), no. 5, 2287–300.

[49] D.S. Horowitz and J.C. Wang, *Torsional rigidity of DNA and length dependence of the free energy of DNA supercoiling*, J. Mol. Biol. **173** (1984), 75–91.

[50] Jim Hoste, Morwen Thistlethwaite, and Jeff Weeks, *The first 1,701,936 knots*, Math. Intelligencer **20** (1998), no. 4, 33–48.

[51] R. Kanaar, A. Klippel, E. Shekhtman, J. M. Dungan, R. Kahmann, and N. R. Cozzarelli, *Processive recombination by the phage Mu Gin system: implications for the mechanisms of DNA strand exchange, DNA site alignment, and enhancer action*, Cell **62** (1990), no. 2, 353–66.

[52] T. Kanenobu and H. Murakami, *Two-bridge knots with unknotting number one.*, Proc. Am. Math. Soc. **98** (1986), 499–502.

[53] V. Katritch, J. Bednar, D. Michoud, J. Dubochet, and A. Stasiak, *Electrophoretic mobility of DNA knots*, Nature **384** (1996), no. 6605, 122.

[54] V. Katritch, J. Bednar, D. Michoud, R. G. Scharein, J. Dubochet, and A. Stasiak, *Geometry and physics of knots*, Nature **384** (1996), 142–145.

[55] P. Kohn, *Two-bridge links with unlinking number one*, Proc. Am. Math. Soc. **113** (1991), 1135–1147.

[56] M. A. Krasnow, A. Stasiak, S. J. Spengler, F. Dean, T. Koller, and N. R. Cozzarelli, *Determination of the absolute handedness of knots and catenanes of DNA*, Nature **304** (1983), no. 5926, 559–60.

[57] P. Kronheimer, T. Mrowka, P. Ozsváth, and Z. Szabó, *Monopoles and lens space surgeries*, Ann. of Math. **165** (2007), 457–546.

[58] B. Laurie, V. Katritch, J. Dubochet, and A. Stasiak, *Geometry and physics of links*, Biophysical Journal **74** (1998).

[59] A.E. Leschziner and N.D.F. Grindley, *The architecture of the $\gamma\delta$ resolvase crossover site synaptic complex revealed by using constrained DNA substrates.*, Mol. Cell **12** (2003), 775–781.

[60] S. D. Levene and H. Tsen, *Analysis of DNA knots and catenanes by agarose-gel electrophoresis*, Methods Mol. Biol. **94** (1999), 75–85.

[61] C. Levine, H. Hiasa, and K.J. Marians, *DNA gyrase and topoisomerase IV: biochemical activities, physiological roles during chromosome replication and drug sensitivities.*, Biochem. Biophys. Acta **1400** (1998), 29–43.

[62] W. Li, S. Kamtekar, Y. Xiong, G. Sarkis, N.D.F. Grindley, and T.A. Steitz, *Structure of a synaptic resolvase tetramer covalently linked to two cleaved DNAs*, Science **309** (2005), 1210–1215.

[63] L.F. Liu and J.L. Davis, *Novel topologically knotted DNA from bacteriophage p4 capsids: studies with topoisomerases*, Nuc. Acids Res. **9** (1981), 3979–3989.

[64] A. Lobb, *A slice genus lower bound from $sl(n)$ khovanov-rozansky homology*, arXiv:math/0702393 (2007).

[65] A. Maxwell and A. D. Bates, *DNA topology*, Oxford University Press, 2005.

[66] S.K. Merickel and R.C. Johnson, *Topological analysis of Hin-catalysed DNA recombination in vivo and in vitro*, Mol. Microbiology **51** (2004), no. 4, 1143–1154.

[67] J.M. Montesinos, *Seifert manifolds that are ramified two-sheeted cyclic coverings*, Bol. Soc. Mat. Mexicana (2) **18** (1973), 1–32.

[68] ———, *Una familia infinita de nudos representados no separables*, Revisita Math. Hisp.-Amer. **33** (1973), no. IV, 32–35.

[69] ———, *Revêtements ramifiés de nœuds, espaces fibrés de seifert et scindements de heegaard*, Publicaciones del seminario Matematico de Galdeano **11** (1984), no. 3.

[70] M. Nollmann, J. He, O. Byron, and W.M. Stark, *Solution structure of the Tn3 resolvase-crossover site synaptic complex*, Mol. Cell **16** (2004), no. 1, 127–137.

[71] P. Ozsváth and Z. Szabó, *Knots with unknotting number one and heegaard floer homology*, Topology **44** (2005), 705–745.

[72] S. Pathania, M. Jayaram, and R.M. Harshey, *Path of DNA within the Mu transpososome. transposase interactions bridging two Mu ends and the enhancer trap five DNA supercoils*, Cell **109** (2002), no. 4, 425–36.

[73] William F. Pohl, *DNA and differential geometry*, Math. Intelligencer **3** (1980/81), no. 1, 20–27. MR617886 (83d:92120)

[74] I.R. Porteous, *Geometric differentiation for the intelligence of curves and surfaces*, Cambridge University Press, Cambridge, England, 2001.

[75] D.E. Pulleyblank, M. Shure, D. Tang, J. Vinograd, and H.-P. Vosberg, *Action of nicking-closing enzyme on supercoiled and nonsupercoiled closed circular DNA: Formation of a Boltzmann distribution of topological isomers*, Proc. Natl. Acad. Sci. USA **72** (1975), no. 11, 4280–4284.

[76] J.A. Rasmussen, *Khovanov homology and the slice genus*, arXiv:math/0402131v1 [math.GT] (2004).

[77] ———, *Khovanov-rozansky homology of two-bridge knots and links.*, Duke Math. J. **136** (2007), 551–583.

[78] P.A. Rice and T.A. Steitz, *Model for a DNA-mediated synaptic complex suggested by crystal packing of $\gamma\delta$ resolvase subunits*, EMBO J **13** (1994), no. 7, 1514–24.

[79] Dale Rolfsen, *Knots and links*, Publish or Perish Press, Berkeley, CA, 1971.

[80] V.V. Rybenkov, C. Ullsperger, A.V. Vologodskii, and N.R. Cozzarelli, *Simplification of DNA topology below equilibrium values by type ii topoisomerases*, Science **277** (1997), no. 5326, 690–3.

[81] H. Schubert, *Über eine numerische Knoteninvariante*, Math. Z. (1954).

[82] N.C. Seeman, *Synthetic single-stranded DNA topology*, Applications of Knot Theory: Proceedings of Symposia in Applied Mathematics (2008).

[83] D. Shore and R.L. Baldwin, *Energetics of DNA twisting: I. Relation between twist and cyclization probability*, J. Mol. Biol. **170** (1983), 957–981.

[84] M. Shure and J. Vinograd, *The number of superhelical turns in native virion SV40 and minicol DNA determined by the band counting method*, Cell **8** (1976), 215–226.

[85] D.S. Silver, *Knot theory's odd origins*, American Scientist **94** (2006), 158–165.

[86] S.J. Spengler, A. Stasiak, and N.R. Cozzarelli, *The stereostructure of knots and catenanes produced by phage λ integrative recombination: implications for mechanism and DNA structure*, Cell **42** (1985), no. 1, 325–34.

[87] M.C. Smith and H.M. Thorpe, *Diversity in the serine recombinases*, Mol Microbiol **44** (2002), no. 2, 299–307.

[88] W.M. Stark, M.R. Boocock, and D.J. Sherratt, *Catalysis by site-specific recombinases*, Trends Genet **8** (1992), 432–439.

[89] A. Stasiak and A. Flamini, *Simulation of action of DNA topoisomerases to investigate boundaries and shapes of spaces of knots*, Biophys. J. **87** (2004), 2968–2975.

[90] D.W. Sumners, C. Ernst, S.J. Spengler, and N.R. Cozzarelli, *Analysis of the mechanism of DNA recombination using tangles*, Quarterly Review of Biophysics **28** (1995), no. 3, 253–313.

[91] O. Sundin and A. Varshavsky, *Terminal stages of SV40 DNA replication proceed via multiply intertwined catenated dimers*, Cell **21** (1980), no. 1, 103–14.

[92] S. Trigueros, J. Arsuaga, M.E. Vazquez, D.W. Sumners, and J. Roca, *Novel display of knotted DNA molecules by two-dimensional gel electrophoresis*, Nucleic Acids Res. **29** (2001), no. 13, E67–7.

[93] I. Torisu, *The determination of the pairs of two-bridge knots or links with gordian distance one*, Proc. Am. Math. Soc. **126** (1998), 1565–1571.

[94] M. Vazquez, S.D. Colloms, and D. Sumners, *Tangle analysis of Xer recombination reveals only three solutions, all consistent with a single three-dimensional topological pathway*, J. Mol. Biol. **346** (2005), no. 2, 493–504.

[95] M. Vazquez and D.W. Sumners, *Tangle analysis of Gin site-specific recombination*, Math. Proc. Cambridge Philos. Soc. **136** (2004), no. 3, 565–582.

[96] A.A. Vetcher, A.Y. Lushnikov, J. Navarra-Madsen, R.G. Scharein, Y.L. Lyubchenko, I.K. Darcy, and S.D. Levene, *DNA topology and geometry in Flp and Cre recombination*, J. Mol. Biol. **357** (2006), 1089–1104.

[97] A.V. Vologodskii, N.J. Crisona, B. Laurie, P. Pieranski, V. Katritch, J. Dubochet, and A. Stasiak, *Sedimentation and electrophoretic migration of DNA knots and catenanes*, J. Mol. Biol. **278** (1998), no. 1, 1–3.

[98] A.V. Vologodskii, *The topology and physics of circular DNA*, CRC Press, Boca Raton, FL, 1992.

[99] A.V. Vologodskii, W. Zhang, V.V. Rybenkov, A.A. Podtelezhnikov, D. Subramanian, J.D. Griffith, and N.R. Cozzarelli, *Mechanism of topology simplification by type ii DNA topoisomerases*, Proc. Natl. Acad. Sci. USA **98** (2001), 3045–3049.

[100] S.A. Wasserman and N.R. Cozzarelli, *Supercoiled DNA-directed knotting by T4 topoisomerase*, J. Biol. Chem. **266** (1991), no. 30, 20567–73.

[101] S.A. Wasserman, J.M. Dungan, and N.R. Cozzarelli, *Discovery of a predicted DNA knot substantiates a model for site-specific recombination*, Science **229** (1985), no. 4709, 171–4.

[102] J.D. Watson and F.H.C. Crick, *A structure for deoxyribose nucleic acid*, Nature **171** (1953), 737–738.

[103] Claude Weber, *Où en sont les mathématiques?*, ch. Questions de topologie en biologie moléculaire. Appendix by Andrzej Stasiak and Jacques Dubochet, pp. 328–347, Vuibert, 2002.

[104] J.H. White and N.R. Cozzarelli, *A simple topological method for describing stereoisomers of DNA catenanes and knots*, Proc. Natl. Acad. Sci. USA **81** (1984), no. 11, 3322–6.

[105] J.H. White, *Self-linking and the Gauss integral in higher dimensions*, Amer. Math. J. **91** (1969), 693–728.

[106] J.H. White, K.C. Millett, and N.R. Cozzarelli, *Description of the topological entanglement of DNA catenanes and knots by a powerful method involving strand passage and recombination*, J. Mol. Biol. **197** (1987), no. 3, 585–603.

[107] E.L. Zechiedrich and N.J. Crisona, *Methods in molecular biology: DNA topoisomerase protocols*, M. Bjornsti and N. Osheroff (eds), vol. 1, ch. Coating DNA with RecA protein to distinguish DNA path by electron microscopy, pp. 99–108, Humana Press, Totowa, NJ, 1989.

DEPARTMENT OF MATHEMATICS, AND CENTRE FOR INTEGRATIVE SYSTEMS BIOLOGY, IMPERIAL COLLEGE LONDON, SOUTH KENSINGTON, LONDON SW7 2AZ, ENGLAND

Proceedings of Symposia in Applied Mathematics
Volume **66**, 2009

Knots and Physics

Louis H. Kauffman

Abstract. This article is an introduction to relationships between knot theory and theoretical physics. We give an exposition of the theory of polynomial invariants of knots and links, the Witten functional integral formulation of knot and link invariants, and the beginnings of topological quantum field theory, and show how the theory of knots is related to a number of key issues in mathematical physics.

1. Introduction

This article is an introduction to some of the relationships between knot theory and theoretical physics. Knots themselves are macroscopic physical phenomena in three-dimensional space, occuring in rope, vines, telephone cords, polymer chains, DNA, certain species of eel and many other places in the natural and man-made world. The study of topological invariants of knots leads to relationships with statistical mechanics and quantum physics. This is a remarkable and deep situation where the study of certain (topological) aspects of the macroscopic world is entwined with theories developed for the subtleties of the microscopic world. The present article is an expanded version of [**27**], an introduction to the mathematical side of these connections, with some hints and references to the related physics.

We begin with a short introduction to knots, links, braids and the bracket polynomial invariant of knots and links. The article then discusses Vassiliev invariants of knots and links, and how these invariants are naturally related to Lie algebras and to Witten's gauge theoretic approach. This part of the article is an introduction to how Vassiliev invariants in knot theory arise naturally in the context of Witten's functional integral.

The article is divided into 7 sections beyond the introduction. Section 2 is a quick introduction to the topology of knots and links. Section 3 discusses Vassiliev invariants and invariants of rigid vertex graphs. Section 4 discusses basics of quantum mechanics, the Schrödinger equation and the Feyman integral approach to quantum amplitudes. Section 5 discusses ordinary integration and gives a guide to the reader for thinking about integrals as equivalence classes of functions so that

2000 *Mathematics Subject Classification.* Primary 57M25, 57M27; Secondary 81T45, 57R56.
Key words and phrases. Knots, links, bracket polynomial, Jones polynomial, Vassiliev invariants, functional integral, topological quantum field theory, quantum computing.

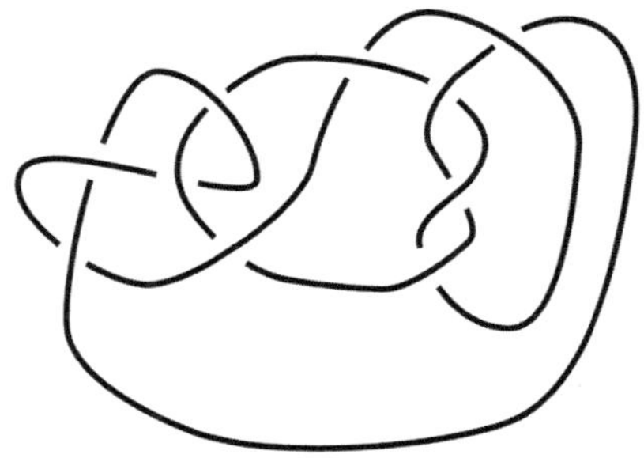

FIGURE 1. **A knot diagram**

f and g are equivalent if their difference has an anti-derivative that vanishes at the boundary conditions of the integral. This point of view is then used in the next section as a way to think about the heuristics of functional integration. Section 6 introduces the basic formalism and shows how Witten's quantum field theoretic functional integral is related directly to Vassiliev invariants. Section 7 discusses the loop transform and loop quantum gravity . Section 8 is an epilogue and guide to the literature about the many ideas, concepts and techniques that we could not touch on in this introductory paper.

The main theme of this paper is to look at the role of Witten's functional integral in studying invariants of knots, links and three-manifolds. We have taken an approach to the Witten integral that is heuristic and uses only advanced calculus. In this sense, our approach is at the same level as Feynman's original approach to his path integral for quantum amplitudes. Accordingly, we give an elementary account of the Feynman path integral in Section 4 and use it to motivate our discussion of the Witten integral in Section 6. For logical reasons we begin with the topology in Sections 2 and 3. Section 3 on Vassiliev invariants allows us to go from the Reidemeister moves for knot diagrams to the intimate relationship of knot invariants with Lie algebras. We meet Lie algebras again in Section 6, and find by the end, that we have arrived at Vassiliev invariants and their weight systems from the vantage of the Witten integral. To make the connections in this order has been our choice. We strongly urge the reader to look at the book by Atiyah [**2**] and Witten's original paper [**41**] for historical perspective, inspiration and many insights that we have not managed to express here.

2. Knots, Braids and Bracket Polynomial

The purpose of this section is to give a quick introduction to the diagrammatic theory of knots, links and braids. A *knot* is an embedding of a circle in three-dimensional space, taken up to ambient isotopy.

Two knots are regarded as equivalent if one embedding can be obtained from the other through a continuous family of embeddings of circles in three-space. A *link* is an embedding of a disjoint collection of circles, taken up to ambient isotopy. Figure 1 illustrates a diagram for a knot. The diagram is regarded both as a schematic picture of the knot, and as a plane graph with extra structure at the

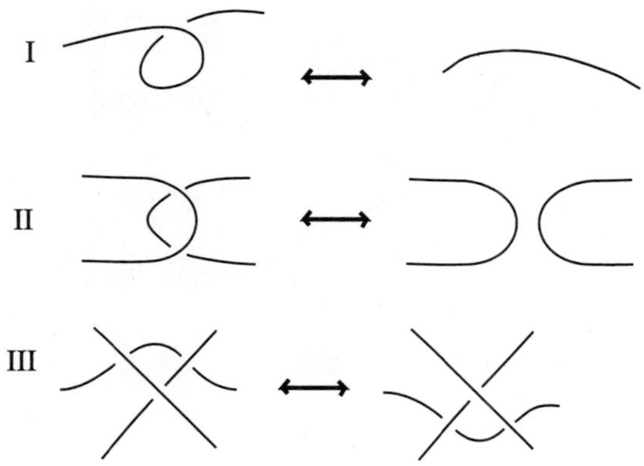

FIGURE 2. **The Reidemeister moves**

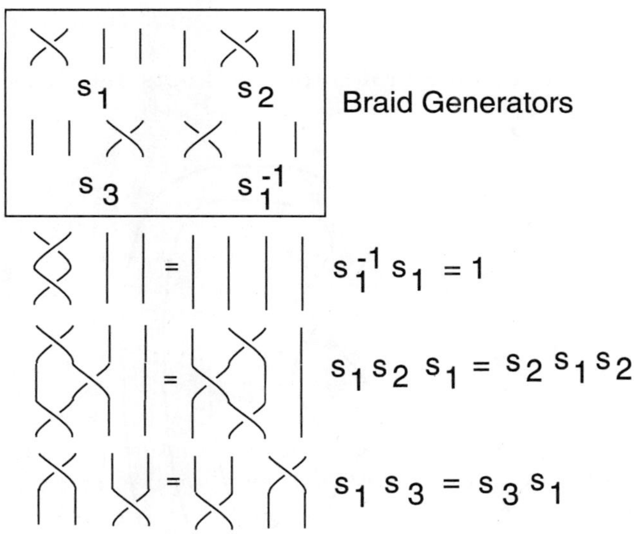

FIGURE 3. **Braid generators**

nodes (indicating how the curve of the knot passes over or under itself by standard pictorial conventions).

Ambient isotopy is mathematically the same as the equivalence relation generated on diagrams by the *Reidemeister moves.* These moves are illustrated in Figure 2. Each move is performed on a local part of the diagram that is topologically identical to the part of the diagram illustrated in this figure (these figures are representative examples of the types of Reidemeister moves) without changing the rest of the diagram. The Reidemeister moves are useful in doing combinatorial topology with knots and links, notably in working out the behaviour of knot invariants. For our purposes, a *knot invariant* is a function defined from knot and

FIGURE 4. **Closing braids to form knots and links**

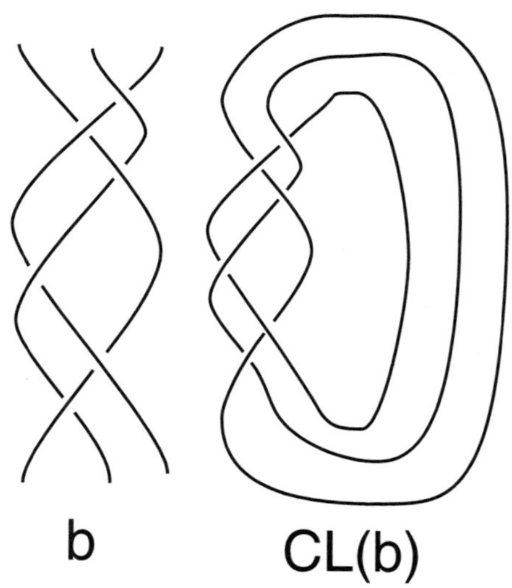

FIGURE 5. **Borromean rings as a braid closure**

link diagrams to some other mathematical object (such as groups or polynomials or numbers) such that equivalent diagrams are mapped to equivalent objects (isomorphic groups, identical polynomials, identical numbers).

Another significant structure related to knots and links is the Artin Braid Group. A *braid* is an embedding of a collection of strands that have their ends in two rows of points that are set one above the other with respect to a choice of vertical. The strands are not individually knotted and they are disjoint from one another. See Figures 3, 4 and 5 for illustrations of braids and moves on braids. Braids can be multiplied by attaching the bottom row of one braid to the top row of the other braid. Taken up to ambient isotopy, fixing the endpoints, the braids form a group under this notion of multiplication. In Figure 3 we illustrate the form of the basic generators of the braid group, and the form of the relations among these generators. Figure 4 illustrates how to close a braid by attaching the top strands to the bottom strands by a collection of parallel arcs. A key theorem of Alexander states that every knot or link can be represented as a closed braid. Thus the theory of braids is critical to the theory of knots and links. Figure 5 illustrates the famous Borromean Rings (a link of three unknotted loops such that any two of the loops are unlinked) as the closure of a braid.

We now discuss a significant example of an invariant of knots and links, the *bracket polynomial.* Two knots or links are said to be *regularly isotopic* if one can be obtained from the other by a sequence of Reidemeister II and III moves. The addition or removal of a curl invvolved in the first Reidemeister move does not occur in regular isotopy. The bracket polynomial is an invariant of regular isotopy, and it can be normalized to produce an invariant of all the Reidemeister moves. This normalized invariant is known as the Jones polynomial [**17**]. The Jones polynomial was originally discovered by a different method than the one given here.

The *bracket polynomial,* $< K > = < K > (A)$, assigns to each unoriented link diagram K a Laurent polynomial in the variable A, such that

(1) If K and K' are regularly isotopic diagrams, then $< K > = < K' >$.

(2) If $K \amalg O$ denotes the disjoint union of K with an extra unknotted and unlinked component O (also called 'loop' or 'simple closed curve' or 'Jordan curve'), then

$$< K \amalg O > = \delta < K >,$$

where

$$\delta = -A^2 - A^{-2},$$

and

$$< O > = 1.$$

(3) $< K >$ satisfies the following formulas

$$< \chi > = A < \asymp > + A^{-1} <)(>$$

$$< \overline{\chi} > = A^{-1} < \asymp > + A <)(>,$$

where the small diagrams represent parts of larger diagrams that are identical except at the site indicated in the bracket. We take the convention that the letter chi, χ, denotes a crossing where *the curved line is crossing over the straight segment.* The barred letter denotes the switch of this crossing, where *the curved line is undercrossing the straight segment.*

In computing the bracket, one finds the following behaviour under Reidemeister move I:

$$< \gamma > = -A^3 < \smile >$$

and

$$< \overline{\gamma} > = -A^{-3} < \smile >$$

where γ denotes a curl of positive type as indicated in Figure 6, and $\overline{\gamma}$ indicates a curl of negative type, as also seen in this figure. The type of a curl is the sign of the crossing when we orient it locally. Our convention of signs is also given in Figure 6.

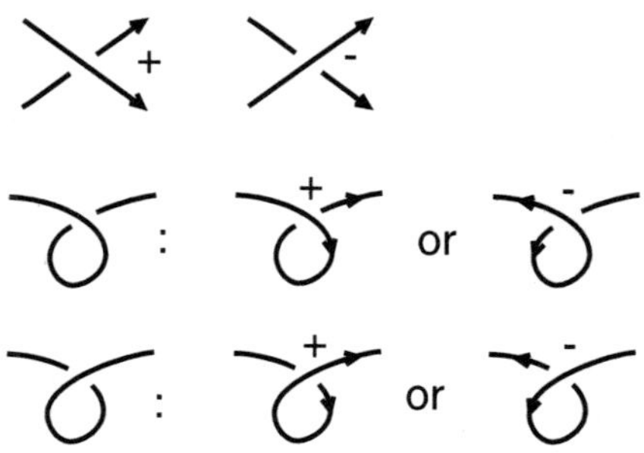

FIGURE 6. **Crossing signs and curls**

Note that the type of a curl does not depend on the orientation we choose. The small arcs on the right hand side of these formulas indicate the removal of the curl from the corresponding diagram.

The bracket is invariant under regular isotopy and can be normalized to an invariant of ambient isotopy by the definition

$$f_K(A) = (-A^3)^{-w(K)} < K > (A),$$

where we chose an orientation for K, and where $w(K)$ is the sum of the crossing signs of the oriented link K. $w(K)$ is called the *writhe* of K. The convention for crossing signs is shown in Figure 6.

The state summation. In order to obtain a closed formula for the bracket, we now describe it as a state summation. Let K be any unoriented link diagram. Define a *state*, S, of K to be a choice of smoothing for each crossing of K. There are two choices for smoothing a given crossing, and thus there are 2^N states of a diagram with N crossings. In a state we label each smoothing with A or A^{-1} as in the expansion formula for the bracket. The label is called a *vertex weight* of the state. There are two evaluations related to a state. The first one is the product of the vertex weights, denoted

$$< K|S > .$$

The second evaluation is the number of loops in the state S, denoted

$$||S||.$$

Define the *state summation*, $< K >$, by the formula

$$< K > = \sum_S < K|S > \delta^{||S||-1}.$$

It follows from this definition that $< K >$ satisfies the equations

$$< \chi > = A < \asymp > + A^{-1} <)(>,$$
$$< K \amalg O > = \delta < K >,$$
$$< O > = 1.$$

The first equation expresses the fact that the entire set of states of a given diagram is the union, with respect to a given crossing, of those states with an A-type smoothing and those with an A^{-1}-type smoothing at that crossing. The second and the third equation are clear from the formula defining the state summation. Hence this state summation produces the bracket polynomial as we have described it at the beginning of the section.

Remark on the Jones polynonmial. By a change of variables one obtains the original Jones polynomial, $V_K(t)$, for oriented knots and links from the normalized bracket:

$$V_K(t) = f_K(t^{-\frac{1}{4}}).$$

By changing variables in this way, and using the bracket state summation, one obtains an oriented state expansion for the Jones polynomial with the basic formulas (δ is the loop value.)

$$\begin{aligned} V_{K_+} &= -t^{1/2}V_{K_0} - tV_{K_\infty}, \\ V_{K_-} &= -t^{-1/2}V_{K_0} - t^{-1}V_{K_\infty}, \\ \delta &= -(t^{1/2} + t^{-1/2}). \end{aligned}$$

Here K_+ and K_-denote oriented positive and negative crossings as shown in Figure 6, K_0 denotes an oriented smoothing at an oriented crossing, and K_∞ denotes the reverse smoothing of such a crossing. We leave verification of this formula as an exercise for the reader. Note that

$$t^{-1}V_{K_+} - tV_{K_-} = (t^{1/2} - t^{-1/2})V_{K_0}.$$

This is the famous skein relation of the Jones polynonmial.

Remark. The bracket polynomial provides a connection between knot theory and physics, in that the state summation expression for it exhibits it as a generalized partition function defined on the knot diagram. Partition functions are ubiquitous in statistical mechanics, where they express the summation over all states of the physical system of probability weighting functions for the individual states. Such physical partition functions contain large amounts of information about the corresponding physical system. Some of this information is directly present in the properties of the function, such as the location of critical points and phase transition. Some of the information can be obtained by differentiating the partition function, or performing other mathematical operations on it.

In fact, by defining a generalization of the bracket polynomial, defined on knot diagrams but not invariant under the Reidemeister moves, we can capture significant partition functions that are physically meaningful. There is not room in this survey to detail how this generalization can be used to express the Potts model for planar graphical configurations, and how it expresses the relationship between the Potts model and the Temperley-Lieb algebra in diagrammatic form. There is much more in this connection with statistical mechanics in that the local weights in a partition function are often expressed in terms of solutions to a matrix equation called the Yang-Baxter equation, that turns out to fit perfectly invariance under the third Reidemeister move. As a result, there are many ways to define partition functions of knot diagrams that give rise to invariants of knots and links. The subject is intertwined with the algebraic structure of Hopf algebras and quantum groups, useful for producing systematic solutions to the Yang-Baxter equation. In fact Hopf algebras are deeply connected with the problem of constructing invariants of three-dimensional manifolds in relation to invariants of knots. We have chosen, in this survey paper, to not discuss the details of these approaches, but rather to proceed to Vassiliev invariants and the relationships with Witten's functional integral. The reader is referred to [**17, 19–21, 36**] for more information about relationships of knot theory with statistical mechanics, Hopf algebras and quantum groups. For topology, the key point is that Lie algebras can be used to construct invariants of knots and links. This is shown nowhere more clearly than in the theory of Vassiliev invariants that we take up in the next section.

3. Vassiliev Invariants and Invariants of Rigid Vertex Graphs

In this section we study the combinatorial topology of Vassiliev invariants. As we shall see, by the end of this section, Vassiliev invariants are directly conncected with Lie algebras, and representations of Lie algebras can be used to construct them. This aspect of link invariants is one of the most fundamental for connections with physics. Just as symmetry considerations in physics lead to a fundamental relationship with Lie algebras, topological invariance leads to a fundamental relationship of the theory of knots and links with Lie algebras.

If $V(K)$ is a (Laurent polynomial valued, or more generally - commutative ring valued) invariant of knots, then it can be naturally extended to an invariant of rigid vertex graphs by defining the invariant of graphs in terms of the knot invariant via an "unfolding" of the vertex. That is, we can regard the vertex as a " black box" and replace it by any tangle of our choice. Rigid vertex motions of the graph preserve the contents of the black box, and hence implicate ambient isotopies of the link obtained by replacing the black box by its contents. Invariants of knots and links that are evaluated on these replacements are then automatically rigid vertex invariants of the corresponding graphs. If we set up a collection of multiple replacements at the vertices with standard conventions for the insertions of the tangles, then a summation over all possible replacements can lead to a graph invariant with new coefficients corresponding to the different replacements. In this way each invariant of knots and links implicates a large collection of graph invariants.

The simplest tangle replacements for a 4-valent vertex are the two crossings, positive and negative, and the oriented smoothing. Let V(K) be any invariant of knots and links. Extend V to the category of rigid vertex embeddings of 4-valent graphs by the formula

$$V(K_*) = aV(K_+) + bV(K_-) + cV(K_0)$$

where K_+ denotes a knot diagram K with a specific choice of positive crossing, K_- denotes a diagram identical to the first with the positive crossing replaced by a negative crossing and K_* denotes a diagram identical to the first with the positive crossing replaced by a graphical node.

There is a rich class of graph invariants that can be studied in this manner. The Vassiliev invariants [**4**] constitute the important special case of these graph invariants where $a = +1$, $b = -1$ and $c = 0$. Thus $V(G)$ is a Vassiliev invariant if

$$V(K_*) = V(K_+) - V(K_-).$$

Call this formula the *exchange identity* for the Vassiliev invariant V. See Figure 7.

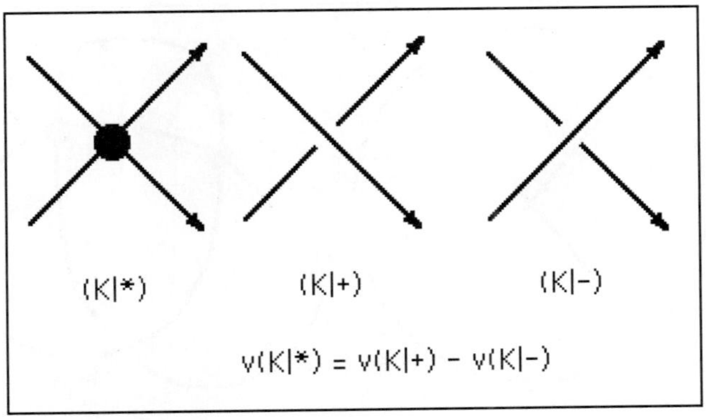

FIGURE 7. **Exchange identity for Vassiliev invariants**

V is said to be of *finite type* k if $V(G) = 0$ whenever $|G| > k$ where $|G|$ denotes the number of (4-valent) nodes in the graph G. The notion of finite type is of extraordinary significance in studying these invariants. One reason for this is the following basic Lemma.

Lemma. If a graph G has exactly k nodes, then the value of a Vassiliev invariant v_k of type k on G, $v_k(G)$, is independent of the embedding of G.

Proof. Apply the exchange identity at a crossing. The difference gives a graph with $k+1$ nodes and hence vanishes. Thus the evaluation of the graph does not depend upon the crossings in the embedding.

The upshot of this lemma is that Vassiliev invariants of type k are intimately involved with certain abstract evaluations of graphs with k nodes. In fact, there are restrictions (the four-term relations) on these evaluations demanded by the topology and it follows from results of Kontsevich [**4**] that such abstract evaluations actually determine the invariants. The knot invariants derived from classical Lie algebras are all built from Vassiliev invariants of finite type. All of this is directly related to Witten's functional integral [**41**].

In the next few figures we illustrate some of these main points. In Figure 8 we show how one associates a so-called chord diagram to represent the abstract graph associated with an embedded graph. The chord diagram is a circle with arcs connecting those points on the circle that are welded to form the corresponding graph. In Figure 9 we illustrate how the four-term relation is a consequence of topological invariance. Figure 9 is the most important part of the connection of the topology and the algebra for Vassiliev invariants.

Look closely at this figure. The top of the figure illustrates a simple topological equivalence related to the rigid vertex. Two segments (parts of a larger diagram) cross to form a vertex. On the left-hand side of the equality a loop passes under all parts of the crossing segments. On the right-hand side of the equality the loop passes over all parts of the crossing segments. The two figures (left and right) are topologically equivalent as they stand, and if they appear inside any larger diagram that has the same local appearance as these figures. The equivalence is obtained by contracting the loop from under the crossing and then expanding it

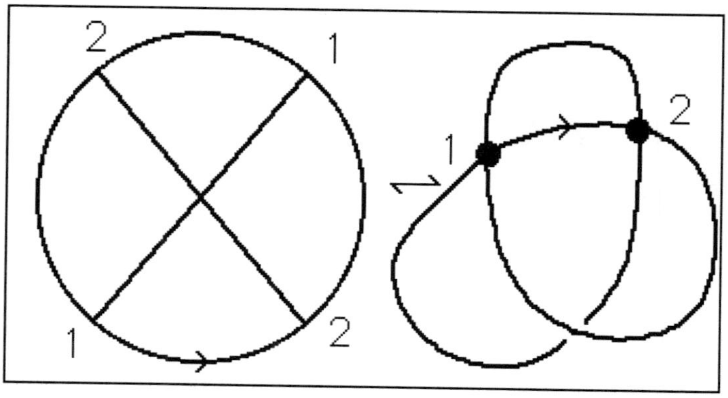

FIGURE 8. **Chord diagrams**

above the crossing, going from the left part of the equality to the right part of the equality. But there is another way to go from the under-crossing diagram to the over-crossing diagram. This second way is given in detail by the four equations that are grouped in the next part of Figure 9. The first equation begins with our familiar left-side diagram and subtracts the result of switching a single crossing. The result (since we are evaluating Vassiliev invariants) is the new diagram on the right with the switched crossing replaced by a node. Thus the diagram on the right has two nodes. We repeat this procedure four times, each time switching one more crossing and arranging the equations so that the last equation involves the right-hand diagram where the loop is entirely above the crossing. Adding these four equations together yields complete cancellation of the sum of their left-sides, and we obtain the boxed equation of diagrams with two nodes each. This is the *embedded four-term relation.* In the embedded four-term relation we see a central node and four neighboring nodes. The neighbors fall into pairs to the left and to the right of the central node. From the embedded four-term relation we go directly to the abstract four-term relation shown directly below it in the chord diagram language. It is a good exercise to make this translation yourself. The abstract four term relation exhibits neighbor relations among crossings in a direct way with one neighbor difference on the left-hand side of the equation, and another neighbor difference on the right-hand side of the equation.

In Figure 10 we show how *the four term relation is a consequence of the abstract pattern of the commutator identity for a matrix Lie algebra.* Here we assume that the Lie algebra is generated by a set of matrices T^a where $a = 1, 2, \ldots, m$. Usually, one takes these matrices to be linearly independent over the ground field (say the complex numbers) for the algebra and assumes that the Lie product is the commutator $[A, B] = AB - BA$ of matrix products. Closure, for this Lie algebra basis, is given by the formula

$$T^aT^b - T^bT^a = \sum_c f_c^{ab}T^c$$

stating that the commutator of the matrices T^a and T^b is equal to a sum of the matrices T^c with coefficients (the structure coefficients of the Lie algebra) f_c^{ab}. Such

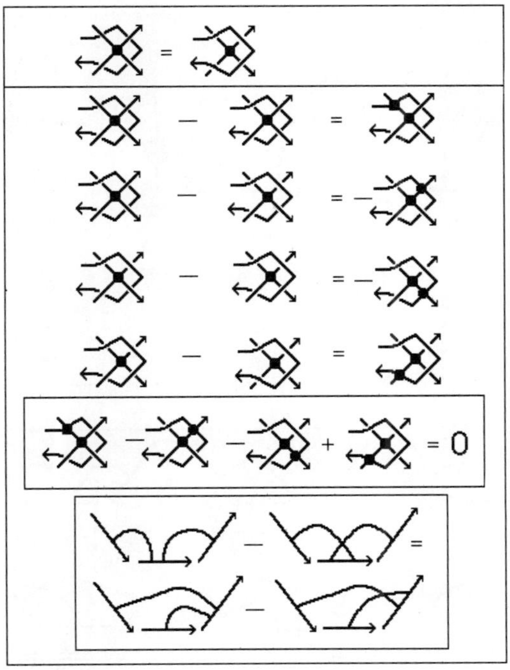

FIGURE 9. **The four term relation from topology**

a relation is the most concrete way to define a matrix Lie algebra. For simplicity, we assume that f_c^{ab} is invariant under cyclic permutation of its indices.

In Figure 10, we show how this closure formula is directly related to the four-term relation. The fit that is illustrated in Figure 10 depends upon enlarging the category of chord diagrams to include trivalent vertices so that this form of the commutator identity (soon to be seen as the Jacobi identity in disquise!) can be expressed. We shall content ourselves with this matrix point of view here, and we will assume that the structure coefficients are invariant under cyclic permutation, as assumption that is not needed in the general case. Note how the derivation at the bottom of Figure 10 works: We take the left-hand side of the four-term relation and see within it the difference between two chord-ends and the transposition of these ends. We interpret this as the pattern of a commutator and replace by the trivalent vertex that represents the structure constants for the Lie algebra. Then this trivalent vertex is seen diagrammatically (by invariance under cyclic permutation) to open up to another difference that is exactly the right-hand side of the four-term relation. In a Lie algebraic category of diagrams, the four-term relation is inevitable. Lie algebra patterns are lurking just beneath the surface of the Reidemeister moves.

There are other levels of abstraction that can be employed here. The same diagrams that we just intepreted as a commutator equation can be interpreted directly in terms of the Jacobi identity that defines a Lie algebra. The four term relation is directly related to a categorical generalisation of Lie algebras. There is

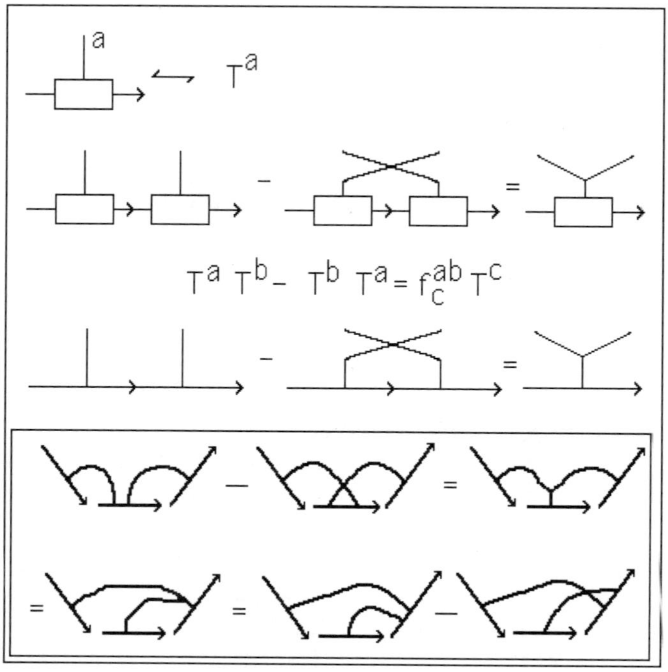

FIGURE 10. **The four term relation from categorical Lie algebra**

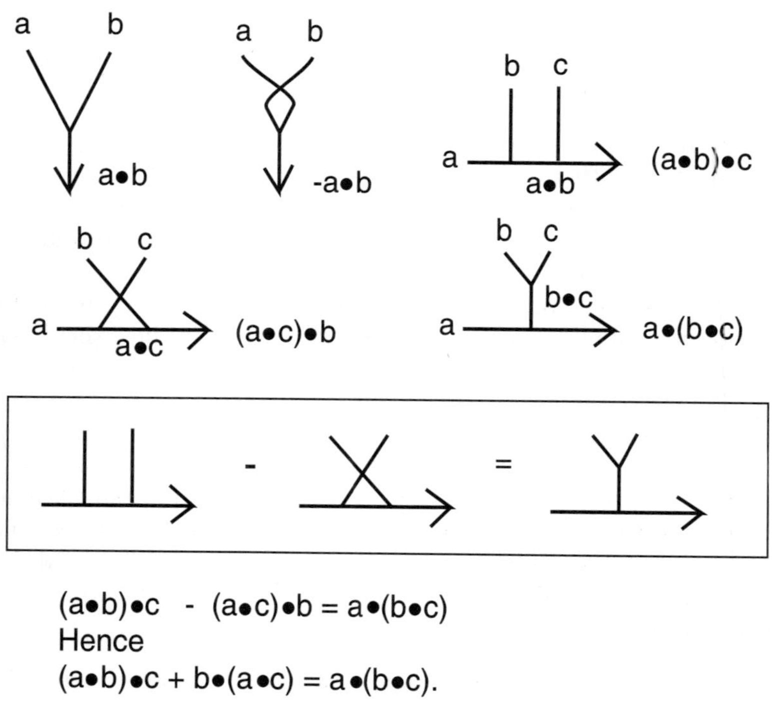

FIGURE 11. **The Jacobi identity**

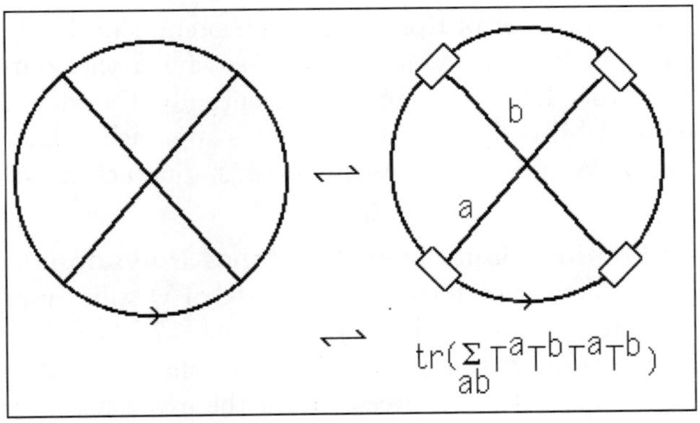

FIGURE 12. **Calculating Lie algebra weights**

insight to be gained from the general definition of a Lie algebra and so we digress in the next paragraph on that point.

A *Lie algebra* is an algebra with a product denoted by $a \bullet b$ (If a and b are in A then then $a \bullet b$ is in A) with the special properties that

$$a \bullet b = -b \bullet a$$

for all a and b in A and

$$a \bullet (b \bullet c) = (a \bullet b) \bullet c + b \bullet (a \bullet c)$$

for all a, b, c in A. For example, if a and b are $n \times n$ matrices then we can define

$$a \bullet b = [a, b] = ab - ba$$

where ab denotes the usual product of matrices, and it is easy to verify that the set of $n \times n$ matrices (say with complex number entries) is a Lie algebra. In Figure 11 we indicate a natural method to diagram products by using a Y shaped graph so that when the top edges of the Y are labeled with a and b then then bottom edge of the Y is labeled with $a \bullet b$. We then see that the diagrammatic relation in the box is exactly the Jacobi identity for a Lie algebra. This is the more fundamental way to see how Lie algebras are involved in the structure of knot theory. The reader will enjoy comparing the diagrams of Figures 10 and 11 to see just how the matrix example and the diagrammatic interpretation of Figure 10 is related to the more general framework and intepretation in Figure 11. It is really remarkable that *the four-term relation is essentially equivalent to the Jacobi identity in a diagrammatic context.*

A *weight system* is an assigment of (numerical) values to chord diagrams so that these numerical values satisfy the four-term relations. (One also needs to rewrite such evaluations such that diagrams with isolated chords are assigned zero, but we will not address this issue here. This is the equivalent of taking care of the first Reidemeister move by a normalization.) Figure 12 illustrates how the weights can be assigned to the chord diagrams by using the Lie algebra: - One inserts Lie algebra matrices into the main circle of the chord diagram and takes a trace of the resulting sum of matrix products. The relationship between Vassiliev invariants and Lie algebras has been known since Bar-Natan's thesis. See also [**22**]. In [**4**] the reader will find a good account of Kontsevich's Theorem, showing how Lie algebra weight systems, and in fact any weight system satsfying the four-term-relation, can be used to construct knot invariants. Conceptually, the ideas in back of the Kontsevich Theorem are directly related to Witten's approach to knot invariants via quantum field theory. We give an exposition of this approach in the next sections of this article.

Example – Vassiliev invariants from the Jones Polynomial. This example shows how the original Jones polynomial is composed of Vassiliev invariants of finite type. Let $V_K(t)$ denote the original Jones polynomial [**17**]. See the remarks at the end of the previous section of this paper. We have the oriented state expansion for the Jones polynomial, obtained directly from the expansion of the normalized bracket polynomial, using $A = t^{-1/4}$. Here K_0 is the locally oriented smoothing of the oriented crossing (K_+ or K_-), and K_∞ is the locally disoriented smoothing.

$$V_{K_+} = -t^{1/2} V_{K_0} - t V_{K_\infty},$$
$$V_{K_-} = -t^{-1/2} V_{K_0} - t^{-1} V_{K_\infty},$$
$$\delta = -(t^{1/2} + t^{-1/2}).$$

Let $t = e^x$. Then

$$V_{K_+} = -e^{x/2} V_{K_0} - e^{x} V_{K_\infty},$$
$$V_{K_-} = -e^{-x/2} V_{K_0} - e^{-x} V_{K_\infty},$$
$$\delta = -(e^{x/2} + e^{-x/2}).$$

Thus

$$V_{K_*} = V_{K_+} - V_{K_-} = -2sinh(x/2) V_{K_0} - 2sinh(x) V_{K_\infty}.$$

Thus x divides V_{K_*}, and therefore x^k divides V_G whenever G is a graph with at least k nodes. Letting

$$V_G(e^x) = \sum_{k=0}^{\infty} v_k(G) x^k,$$

we see that this condition implies that $v_k(G)$ vanishes whenever G has more than k nodes. Hence *the coefficients of the powers of x in the expansion of $V_K(e^x)$ are Vassiliev invariants of finite type!* This result was first observed by Birman and Lin [**6**] by a different argument.

4. Schrödinger's Equation and the Feynman Path Integral

4.1. A quick review of quantum mechanics. To recall principles of quantum mechanics it is useful to have a quick historical recapitulation. Quantum mechanics really began when DeBroglie introduced the fantastic hypothesis that matter (such as an electron) is accompanied by a wave that guides its motion and produces interference phenomena just like the waves on the surface of the ocean or the diffraction effects of light going through a small aperture. DeBroglie's idea was successful in explaining the properties of atomic spectra. In this domain, his wave hypothesis led to the correct orbits and spectra of atoms, formally solving a puzzle that had been only described in ad hoc terms by the preceding theory of Niels Bohr. In Bohr's theory of the atom, the electrons are restricted to move only in certain elliptical orbits. These restrictions are placed in the theory to get agreement with the known atomic spectra, and to avoid a paradox! The paradox arises if one thinks of the electron as a classical particle orbiting the nucleus of the atom. Such a particle is undergoing acceleration in order to move in its orbit. Accelerated charged particles emit radiation. Therefore the electron should radiate away its energy and spiral into the nucleus! Bohr commanded the electron to only occupy certain orbits and thereby avoided the spiral death of the atom - at the expense of logical consistency.

DeBroglie hypothesized a wave associated with the electron and he said that an integral multiple of the length of this wave must match the circumference of the electron orbit. Thus, not all orbits are possible, only those where the wave pattern can "bite its own tail". The mathematics works out, providing an alternative to Bohr's picture. DeBroglie had waves, but he did not have an equation describing the spatial distribution and temporal evolution of these waves. Such an equation was discovered by Erwin Schrödinger. Schrödinger relied on inspired guesswork, based on DeBroglie's hypothesis and produced a wave equation, known ever since as the Schrödinger equation. Schrödinger's equation was enormously successful, predicting fine structure of the spectrum of hydrogen and many other aspects of physics. Suddenly a new physics, quantum mechanics, was born from this musical hypothesis of DeBroglie.

Along with the successes of quantum mechanics came a host of extraordinary problems of interpretation. What is the status of this wave function of Schrödinger and DeBroglie. Does it connote a new element of physical reality? Is matter "nothing but" the patterning of waves in a continuum? How can the electron be a wave and still have the capacity to instantiate a very specific event at one place and one time (such as causing a bit of phosphor to glow there on your television screen)? It came to pass that Max Born developed a statistical interpretation of the wave function wherein the wave determines a probability (frequency of occurrence) for the appearance of the localized particulate phenomenon that one wanted to call an "electron". In this story the wavefunction ψ takes values in the complex numbers and the associated probability is $\psi^*\psi$, where ψ^* denotes the complex conjugate of ψ. Mathematically, this is a satisfactory recipe for dealing with the theory, but it leads to further questions about the exact character of the statistics. If quantum theory is inherently statistical, then it can give no complete information about the motion of the electron. In fact, there may be no such complete information available even in principle. Electrons manifest as particles when they are observed in a certain manner and as waves when they are observed in another, complementary manner. This is a capsule summary of the view taken by Bohr,Heisenberg and Born. Others, including DeBroglie, Einstein and Schrödinger, hoped for a more direct and deterministic theory of nature. As we shall see in the following sections, the statistical nature of quantum theory has a formal side that can be exploited to understand the topological properties of such mundane objects as knotted ropes in space and spaces constructed by identifying the sides of polyhedra. These topological applications of quantum mechanical ideas are exciting in their own right. They may shed light on the nature of quantum theory itself.

In this section we review a bit of the mathematics of quantum theory. Recall the equation for a wave:

$$f(x,t) = sin((2\pi/l)(x - ct)).$$

With x interpreted as the position and t and as the time, this function describes a sinusoidal wave traveling with velocity c. We define the wave number $k = 2\pi/l$ and the frequency $\omega = (2\pi c/l)$ where l is the wavelength. Thus we can write

$$f(x,t) = sin(kx - wt).$$

Note that the velocity, c, of the wave is given by the ratio of frequency to wave number, $c = \omega/k$.

DeBroglie and Einstein hypothesized two fundamental relationships: between energy and frequency, and between momentum and wave number. These relationships are summarized in the equations

$$E = \hbar\omega,$$

$$p = \hbar k,$$

where E denotes the energy associated with a wave and p denotes the momentum associated with the wave. Here $\hbar = h/2\pi$ where h is Planck's constant. For DeBroglie the discrete energy levels of the orbits of electrons in an atom of hydrogen could be explained by restrictions on the vibrational modes of waves associated with the motion of the electron. His choices for the energy and the momentum in relation to a wave are not arbitrary. They are designed to be consistent with the notion that the wave or wave packet moves along with the electron. That is, the velocity of the wave packet is designed to be the velocity of the "corresponding" material particle. It is worth illustrating how DeBroglie's idea works. Consider two waves whose frequencies are very nearly the same. If we superimpose them (as a piano tuner superimposes his tuning fork with the vibration of the piano string), then there will be a new wave produced by the interference of the original waves. This new wave pattern will move at its own velocity, different (and generally smaller) than the velocity of the original waves. To be specific, let

$$f(x,t) = sin(kx - \omega t)$$

and

$$g(x,t) = sin(k'x - \omega' t).$$

Let

$$h(x,t) = sin(kx - \omega t) + sin(k'x - \omega' t) = f(x,t) + g(x,t).$$

A little trigonometry shows that

$$h(x,t) = cos(((k - k')/2)x - ((\omega - \omega')/2)t)sin(((k + k')/2)x - ((\omega + \omega')/2)t).$$

If we assume that k and k' are very close and that ω and ω' are very close, then $(k + k')/2$ is approximately k, and $(\omega + \omega')/2$ is approximately ω. Thus $h(x,t)$ can be represented by

$$H(x.t) = cos((dk/2)x - (d\omega/2)t)f(x,t)$$

where $dk = (k - k')/2$ and $d\omega = (\omega - \omega')/2$. This means that the superposition, $H(x,t)$, behaves as the waveform $f(x,t)$ carrying a slower-moving "wave packet"

$$G(x,t) = cos((dk/2)x - (d\omega/2)t).$$

Since the wave packet has the equation given above, we see that that the velocity of this wave packet is $v_g = d\omega/dk$. Recall that wave velocity is the ratio of frequency to wave number. Now according to DeBroglie, $E = \hbar\omega$ and $p = \hbar k$, where E and p are the energy and momentum associated with this wave packet. Thus we get the formula $v_g = dE/dp$. In other words, the velocity of the wave packet is the rate of change of its energy with respect to its momentum. Now this is exactly in accord with the well-known classical laws for a material particle! For such a particle,

$$E = mv^2/2$$

and

$$p = mv.$$

Thus

$$E = p^2/2m$$

and

$$dE/dp = p/m = v.$$

It is this astonishing concordance between the simple wave model and the classical notions of energy and momentum that initiated the beginnings of quantum theory.

Schrödinger's equation. Schrödinger answered the question: What is the wave equation for DeBroglie's waves? Writing an elementary wave in complex form

$$\psi = \psi(x,t) = exp(i(kx - \omega t)),$$

we see that we can extract DeBroglie's energy and momentum by differentiating:

$$i\hbar\partial\psi/\partial t = Ey$$

and

$$-i\hbar\partial\psi/\partial x = p\psi.$$

This led Schrödinger to postulate the identification of dynamical variables with operators so that the first equation,

$$i\hbar\partial\psi/\partial t = E\psi,$$

is promoted to the status of an equation of motion while the second equation becomes the definition of momentum as an operator:

$$p = -i\hbar\partial/\partial x.$$

Once p is identified as an operator, the numerical value of momentum is associated with an eigenvalue of this operator, just as in the example above. In our example $p\psi = \hbar k\psi$. In this formulation, the position operator is just multiplication by x itself. Once we have fixed specific operators for position and momentum, the operators for other physical quantities can be expressed in terms of them. We obtain the energy operator by substitution of the momentum operator in the classical formula for the energy:

$$E = p^2/2m + V,$$
$$E = -(\hbar^2/2m)\partial^2/\partial x^2 + V.$$

Here V is the potential energy, and its corresponding operator depends upon the details of the application. With this operator identification for E, Schrödinger's equation

$$i\hbar\partial\psi/\partial t = -(\hbar^2/2m)\partial^2\psi/\partial x^2 + V\psi$$

is an equation in the first derivatives of time and in second derivatives of space. In this form of the theory one considers general solutions to the differential equation and this in turn leads to excellent results in a myriad of applications. In quantum theory, observation is modelled by the concept of eigenvalues for corresponding operators. The quantum model of an observation is a projection of the wave function into an eigenstate. An energy spectrum $\{E_k\}$ corresponds to wave functions y satisfying the Schrödinger equation, such that there are constants E_k with $Ey = E_k y$. An observable (such as energy) E is a Hermitian operator on a Hilbert space of wavefunctions. Since Hermitian operators have real eigenvalues, this provides the link with measurement for the quantum theory.

It is important to notice that there is no mechanism postulated in this theory for how a wave function is sent into an eigenstate by an observable. Just as mathematical logic need not demand causality behind an implication between propositions, the logic of quantum mechanics does not demand a specified cause behind an observation. This absence of an assumption of causality in logic does not obviate the possibility of causality in the world. Similarly, the absence of causality in quantum observation does not obviate causality in the physical world. Nevertheless, the debate over the interpretation of quantum theory has often led its participants into asserting that causality has been demolished in physics. Note that the operators for position and momentum satisfy the equation $xp - px = \hbar i$. This corresponds directly to the equation obtained by Heisenberg, on other grounds, that dynamical variables can no longer necessarily commute with one another. In this way, the points of view of DeBroglie, Schrödinger and Heisenberg came together, and quantum mechanics was born. In the course of this development, interpretations varied widely. Eventually, physicists came to regard the wave function not as a generalised wave packet, but as a carrier of information about possible observations. In this way of thinking $\psi^*\psi$ (ψ^* denotes the complex conjugate of ψ.) represents the probability density for finding the "particle" (a particle is an observable with local spatial characteristics) in a given region in space (at a given time). The mathematical fact that bolsters this point of view is that it follows from the Schrödinger equation that $\int_{-\infty}^{\infty} \psi^*\psi dx$ is time independent. This means that one can normalize $\psi(x,t)$ so that $\int_{-\infty}^{\infty} \psi^*\psi dx = 1$, and then indeed $\psi^*\psi$ has the form of a probability density. Furthermore, this constancy of the space integral of $\psi^*\psi$ can be interpreted as saying that the time-evolution of the wave function is *unitary* in the sense that it preserves the norm $||\psi|| = \sqrt{\int_{-\infty}^{\infty} \psi^*\psi dx}$.

It is worth noting that finite dimensional approximations to quantum mechanics can be studied by using unitary matrices over the complex numbers to represent the physcial processes that we have here assigned to the temporal evolution of the Schrödinger equation. Such finite dimensional versions of quantum mechanics are central to the formulation of algorithms for quantum computation. There is a deep and continuing interaction between knot theory, topology and quantum computation and quantum information theory. We do not have space in this paper to treat that subject, but the reader may enjoy looking at [**16**, **24**–**26**].

4.2. The Feynman path integral. In his 1942 PhD Thesis and subsequent research [**11**–**14**] Richard Feynman discovered and articulated a new approach to quantum mechanics based on a process of integration over all possible paths that a particle can take from one point in space to another. To make this idea specific, suppose that $\psi(x,t)$ denotes the wave function (amplitude) for finding a particle at the point x and time t, given that the particle is known to have started at

$x = 0$ at time $t = 0$. We can regard the latter as an initial condition or boundary condition and attempt to solve Schrödinger's equation for the later values of the wave function. Feynman proceeds by considering each path that the particle can take (with no regard for the constraints of physical law) and assigns to each path a complex number, namely

$$e^{\frac{i}{\hbar}\int_\alpha L dt}$$

where $L = m\dot{x}^2/2 - V(x)$ (with $\dot{x} = dx/dt$) is the Lagrangian – the difference between the kinetic and potential energy at a given location in the path – and $\int_\alpha L dt$ denotes the integral of the Langrangian L along the path α from time $t = 0$ to the final time t. Thus for each path α there is a phase (unit complex number) contribution. And Feynman says that the amplitude will be given by summing these phases over all the paths. In formalism this reads

$$\psi(x,t) = \int_{\{\alpha|\alpha(0)=0,\alpha(t)=x\}} e^{\frac{i}{\hbar}\int_\alpha L dt} dP.$$

This is the Feynman path integral formulation of quantum mechanics.

Strictly speaking, one does not actually know the measure on the space of all possible paths that is denoted formally here by the symbol dP. Nevertheless, this integration process makes sense in relation to the principles of quatnum mechanics and so can be investigated by summing over and taking limits for certain path sums that are appropriate for a given physical context. Feyman understood this and wrote "The neccessity to redefine the method of integration does not destroy the concept of integration. So we feel that the possible awkwardness of the special definition of the sum over all paths may eventually require new definitions to be formulated. Nevertheless, the concept of the sum over all paths, like the concept of an ordinary integral, is independent of a special definition and valid in spite of the failure of such definitions." **[13]** (p. 34). The subject of measure theories for path integrals is a continuing topic of research, and at the same time the concept of the quantum amplitude as a sum over paths has become fundamental to calculation and understanding in quantum mechanics.

There are many pointers to the correct nature of Feynman's approach via path integration. Perhaps the most convincing evidence is Feynman's original calculation showing that the path integration process yields a solution to Schrödinger's equation. His calculation goes as follows: Let ϵ be a small increment of time. Then the path integral formula tells us that

$$\psi(x, t+\epsilon) = \int e^{\frac{i\epsilon}{\hbar}[\frac{m}{2}(\frac{x-y}{\epsilon})^2 - V(x)]}\psi(y,t)\frac{dy}{A}$$

for an appropriate constant A. Letting $y = \eta + x$, we have

$$\psi(x, t+\epsilon) = \int e^{\frac{i}{\hbar}[\frac{m\eta^2}{2\epsilon} - \epsilon V(x)]}\psi(x+\eta,t)\frac{d\eta}{A}$$

He then argues that "only values of η close to zero will contribute to the integral, because, for small ϵ, other values of η make the exponential oscillate so rapidly that there will arise little contribution to the integral." **[15]** (p. 29). Then, taking η small, he expands $\psi(x, t+\epsilon)$ in a Taylor series around $\eta = 0$, uses the values of Gaussian integrals such as

$$\int_{-\infty}^{\infty} e^{-Kx^2} dx = \sqrt{\frac{\pi}{K}},$$

and

$$\int_{-\infty}^{\infty} x^2 e^{-Kx^2} dx = \frac{1}{2K}\sqrt{\frac{\pi}{K}}.$$

and finds Schrödinger's equation in the form

$$\psi(x,t) + \epsilon\partial\psi(x,t)/\partial t = \psi(x,t) - \frac{i\epsilon}{\hbar}V(x)\psi(x,t) + \frac{\hbar i\epsilon}{2m}\partial^2\psi/\partial x^2.$$

See [**15**] (pp. 29-30). We give more details in the next paragraph.

The Taylor expansion is

$$\psi(x, t+\epsilon) = \frac{e^{\frac{-i\epsilon V(x)}{\hbar}}}{A}\int e^{\frac{im\eta^2}{\hbar 2\epsilon}}[\psi(x,t) + \eta\frac{\partial\psi(x,t)}{\partial x} + \frac{\eta^2}{2}\frac{\partial^2\psi(x,t)}{\partial x^2} + \cdots]d\eta.$$

Now use the Gaussian integrals

$$\int_{-\infty}^{\infty} e^{\frac{im\eta^2}{\hbar 2\epsilon}} d\eta = \sqrt{\frac{2\pi\hbar\epsilon i}{m}},$$

and

$$\int_{-\infty}^{\infty} \eta^2 e^{\frac{im\eta^2}{\hbar 2\epsilon}} d\eta = \sqrt{\frac{2\pi\hbar\epsilon i}{m}}\frac{\hbar\epsilon i}{m}.$$

This rewrites the Taylor series as follows.

$$\psi(x, t+\epsilon) = \frac{\sqrt{\frac{2\pi\hbar\epsilon i}{m}}}{A}e^{\frac{-i\epsilon V(x)}{\hbar}}[\psi(x,t) + \frac{\hbar\epsilon i}{2m}\frac{\partial^2\psi}{\partial x^2} + O(x^2)].$$

Taking

$$A(\epsilon) = \sqrt{\frac{2\pi\hbar\epsilon i}{m}},$$

we get

$$\psi(x,t) + \epsilon\partial\psi(x,t)/\partial t = \psi(x,t) - \frac{i\epsilon}{\hbar}V(x)\psi(x,t) + \frac{\hbar i\epsilon}{2m}\partial^2\psi/\partial x^2.$$

Hence $\psi(x,t)$ satisfies the Schrödinger equation.

It is very rewarding to compare the two paths that we have sketched here that lead to the Schrödinger equation. The first path begins with DeBroglie's theory of matter waves and Schrödinger's understanding that it is possible to extract physical quantities from wave functions by using differential operators. The Feynman path does not explicitly use differential operators in this way, but arrives at the wave equation from its solution! Furthermore one can see from Feynman's formulation how the classical physics is related to the quantum physics, for the main contribution to the Feynman integral should be when the variation of the action integral $S(\alpha) = \int_\alpha Ldt$ (with respect to the path α) is zero. For such paths, the nearby paths will all contribute essentially the same phase, and these paths will make a major contribution to the Feynman integral. But it was well-known since Lagrange, that those paths for which the variation of the action integral vanishes are exactly the classical paths, the solutions to Newton's equations! Thus the Feynman path integral gives a new context to the classical principle of least action and shows that quantum mechanics is an extension of the principle of least action into a new domain.

All these themes of Feynman's path integral turn out, in the hands of Edward Witten, to inform the topology of knots, links and three-manifolds. It is the purpose of the next section of this paper, to supply some tools for thinking about integration, and then we shall look at the quantum field theoretic path integral of Witten that is so closely related to knot invariants, the Jones polynomial and the Vassiliev invariants. Just as with Feynman's original integral, the problems of finding the right measure theory keep this theory from being completely rigorous. We shall use elementary heuristics, reminiscent of Feynman's original approach to his integral, to see some of the properties of the Witten integral.

5. Integration without Integration

In this section we discuss integration as a preparation for the next section. In that section we will look at Witten's quantum field theoretic approach to knot and link invariants via a functional integral. In this section we point out that much of what can be done in integration can be done without knowing whether the integral exists! How can this be? Remember elementary calculus where the students are told by their textbooks that an integral is an anti-derivative. This means that "up to constants" $\int f = \int g$ means that $f - g = Dh$ where D is the derivative operator under discussion. In this sense an integral is an equivalence class of fuctions. And we shall see in the next section that by looking at the structure of equivalence classes of certain functions of gauge fields, topology appears! But we are getting ahead of the story. Now for some calculus.

Recall that if $Z = \int_{-\infty}^{\infty} e^{-x^2/2}dx$ then

$$Z^2 = \int_{-\infty}^{\infty}\int_{-\infty}^{\infty} e^{-(x^2+y^2)/2}dxdy = \int_0^{2\pi}\int_0^{\infty} e^{-r^2/2}rdrd\theta$$

$$= 2\pi \int_0^{\infty} e^{-r^2/2}rdr = 2\pi.$$

Whence

$$Z = \sqrt{2\pi}.$$

Furthermore, if

$$Z(J) = \int_{-\infty}^{\infty} e^{-x^2/2+Jx}dx,$$

then

$$\begin{aligned} Z(J) &= \int_{-\infty}^{\infty} e^{-(x-J)^2/2+J^2/2}dx \\ &= e^{J^2/2}\int_{-\infty}^{\infty} e^{-(x-J)^2/2}dx \\ &= e^{J^2/2}\int_{-\infty}^{\infty} e^{-x^2/2}dx \\ &= e^{J^2/2}Z(0) = \sqrt{2\pi}e^{J^2/2}. \end{aligned}$$

Now examine how much of this calculation could be done if we did not know about the existence of the integral, or if we did not know how to calculate explicitly the values of these integrals across the entire real line. Given that we believed in the existence of the integrals, and that we could use properties such as change of variable giving

$$\int_{-\infty}^{\infty} e^{-(x-J)^2/2}dx = \int_{-\infty}^{\infty} e^{-x^2/2}dx,$$

we could deduce the relative result stating that

$$Z(J) = \int_{-\infty}^{\infty} e^{-x^2/2+Jx}dx = e^{J^2/2}\int_{-\infty}^{\infty} e^{-x^2/2}dx.$$

From this we can deduce that

$$d^nZ(J)/dJ^n|_{J=0} = d^n/dJ^n\int_{-\infty}^{\infty} e^{-x^2/2+Jx}dx = \int_{-\infty}^{\infty} x^n e^{-x^2/2}dx.$$

Hence

$$\int_{-\infty}^{\infty} x^n e^{-x^2/2}dx = d^n(e^{J^2/2})/dJ^n|_{J=0}\int_{-\infty}^{\infty} e^{-x^2/2}dx.$$

But now, lets go a step further and imagine that we really have no theory of integration available. Then we are in the position of freshman calculus where one defines $\int f$ to be "any" function g such that $dg/dx = f$. One *defines* the integral in this form of elementary calculus to be the anti-derivative, and this takes care of the matter for a while! What are we really doing in freshman calculus? We are noting that for integration on an interval $[a,b]$, if two functions f and g satisfy $f - g = dh/dx$ for some differentiable function h, then we have that

$$\int_a^b (f-g) = \int_a^b dh/dx = h(b) - h(a).$$

If the function $h(x)$ vanishes as x goes to infinity, then we have that

$$\int_{-\infty}^{\infty} f dx = \int_{-\infty}^{\infty} g dx$$

when $f - g = dh/dx$. This suggests turning things upside down and *defining an equivalence relation on functions*

$$f \sim g$$

if

$$f - g = dh/dx$$

where $h(x)$ is a function vanishing at infinity. Then *we define the integral*

$$\int f(x)$$

to be the equivalence class of the function $f(x)$. This "integral" represents integration from minus infinity to plus infinity but it is defined only as an equivalence class of functions. An "actual" integral, like the Riemann, Lesbeque or Henstock integral is a well-defined real valued function that is constant on these equivalence classes.

We shall say that $f(x)$ is *rapidly vanishing at infinity* if $f(x)$ and all its derivatives are vanishing at infinity. For simplicity, we shall assume that all functions under consideration have convergent power series expansions so that

$$f(x+J) = f(x) + f'(x)J + f''(x)J^2/2! + \cdots,$$

and that they are rapidly vanishing at infinity. It then follows that

$$f(x+J) = f(x) + \frac{d}{dx}[f(x)J + f'(x)J^2/2! + \cdots] \sim f(x),$$

and hence we have that $\int f(x+J) = \int f(x)$, giving translation invariance when J is a constant.

We have shown the following Proposition.

Proposition. Let $f(x), g(x), h(x)$ be functions rapidly vanishing at infinity (with power series representations). Let $\int f$ denote the equivalence class of the function f where $f \sim g$ means that $f - g = Dh$ where $Dh = dh/dx$. Then this integral satisfies the following properties

(1) If $f \sim g$ then $\int f = \int g$.
(2) If k is a constant, then $\int (kf + g) = k \int f + \int g$.
(3) If J is a constant, then $\int f(x+J) = \int f(x)$.
(4) $\int Dh = 0$ where 0 denotes the equivalence class of the zero function. Hence $\int f(Dg) + \int (Df)g = \int D(fg) = 0$, so that integration by parts is valid with vanishing boundary conditions at infinity.

Note that $e^{-x^2/2}$ is rapidly vanishing at infinity. We now see that most of the calculations that we made about $e^{-x^2/2}$ were actually statements about the equivalence class of this function:

$$e^{-x^2/2+Jx} = e^{-(x-J)^2/2+J^2/2} = e^{J^2/2}e^{-(x-J)^2/2} \sim e^{J^2/2}e^{-x^2/2},$$

whence

$$\int e^{-x^2/2+Jx} = e^{J^2/2}\int e^{-x^2/2}.$$

5.1. Functional derivatives. In order to generalize the ideas presented in this section to the context of functional integrals, we need to discuss the concept of functional derivatives. We are given a *functional* $F(\alpha(x))$ whose argument $\alpha(x)$ is a function of a variable x. We wish to define the *functional derivative* $\delta F(\alpha(x))/\delta\alpha(x_0)$ of $F(\alpha(x))$ with respect to $\alpha(x)$ at a given point x_0. The idea is to regard each $\alpha(x_0)$ as a separate variable, giving $F(\alpha(x))$ the appearance of a function of infinitely many variables. In order to formalize this notion one needs to use generalized functions (distributions) such as the Dirac delta function $\delta(x)$, a distribution with the property that $\int_a^b \delta(x_0)f(x)dx = f(x_0)$ for any integrable function $f(x)$ and point x_0 in the interval $[a,b]$. One defines the functional derivative by the formula

$$\delta F(\alpha(x))/\delta\alpha(x_0) = lim_{\epsilon\to 0}[F(\alpha(x)+\delta(x_0)\epsilon) - F(\alpha(x))]/\epsilon.$$

Note that if

$$F(\alpha(x)) = \alpha(x)^2$$

then

$$\begin{aligned}\delta F(\alpha(x))/\delta\alpha(x_0) &= lim_{\epsilon\to 0}[(\alpha(x)+\delta(x_0)\epsilon)^2 - \alpha(x)^2]/\epsilon \\ &= lim_{\epsilon\to 0}[2\alpha(x)\delta(x_0)\epsilon + \delta(x_0)^2\epsilon^2]/\epsilon \\ &= 2\alpha(x)\delta(x_0).\end{aligned}$$

While if

$$G(\alpha(x)) = \int_a^b \alpha(x)^2 dx$$

then

$$\delta G(\alpha(x))/\delta\alpha(x_0) = 2\alpha(x_0)$$

when $x_0 \in [a,b]$. More generally, if

$$G(\alpha(x)) = \int_a^b f(\alpha(x))dx$$

for a differentiable function f, then

$$\delta G(\alpha(x))/\delta\alpha(x_0) = f'(\alpha(x_0)).$$

These examples show that the results of a functional differentiation can be either a distribution or a function, depending upon the context of the original functional.

In the case of a path integral of the type used in quantum mechanics, one wants to integrate a functional $F(p)$ over paths $p(t)$ with t in an interval $[0,1]$. The functional takes the form

$$F(p) = e^{(i/\hbar)\int_0^1 L(p(t))dt}$$

and the traditional Feynman path integral has the form

$$\int dPe^{(i/\hbar)\int_0^1 L(p(t))dt},$$

giving the amplitude for a particle to travel from $a = p(0)$ to $b = p(1)$, the integration proceeding over all paths with these initial and ending points.

Here the equivalence relation corresponding to the functional integral is $F \sim G$ if $F - G = DH$ where

$$DH = \delta H(p)/\delta p(t_0)$$

for some time t_0 and some $H(p)$. Again we need to specify the class of functionals and to say what it means for a functional to "vanish at infinity." Since we are integrating over all paths, we need a notion of size for a path. This can be defined by

$$||p|| = (\int_0^1 |p(t)|^2 dt)^{1/2}.$$

Note that for $F(p) = e^{(i/\hbar)\int_0^1 L(p(t))dt}$ we have

$$\delta F(p)/\delta p(t_0) = (i/\hbar)[\delta \int_0^1 L(p(t))dt/\delta p(t_0)]F(p)$$
$$= (i/\hbar)L'(p(t_0))F(p).$$

Here we see the fact that the integral can be dominated by contributions from paths where this variation is zero. Note that in order to estimate this stationary phase contribution to the functional integral, one needs more than just a definition of the integral as an equivalence class of functionals. Nevertheless, we shall see in the next section that these equivalence classes do give insight into the topology associated with Witten's integral.

6. Vassiliev Invariants and Witten's Functional Integral

In [**41**] Edward Witten proposed a formulation of a class of 3-manifold invariants as generalized Feynman integrals taking the form $Z(M)$ where

$$Z(M) = \int DAe^{(ik/4\pi)S(M,A)}.$$

Here M denotes a 3-manifold without boundary and A is a gauge field (also called a gauge potential or gauge connection) defined on M. The gauge field is a one-form on a trivial G-bundle over M with values in a representation of the Lie algebra of G. The group G corresponding to this Lie algebra is said to be the gauge group. In this integral the action $S(M,A)$ is taken to be the integral over M of the trace of the Chern-Simons three-form $A \wedge dA + (2/3)A \wedge A \wedge A$. (The product is the wedge product of differential forms.) We shall refer to

$$S(M,A) = \int_{M^3} trace(A \wedge dA + (2/3)A \wedge A \wedge A)$$

as the *Chern-Simons action*. $Z(M)$ integrates over all gauge fields modulo gauge equivalence. The trace of the Chern-Simons three-form takes the role of the Lagrangian in this theory.

The formalism and internal logic of the Witten integral supports the existence of a large class of topological invariants of 3-manifolds, and associated invariants of knots and links in these manifolds. The invariants associated with this integral have been given rigorous combinatorial descriptions, yet questions and conjectures arising from the integral formulation are still outstanding. Specific conjectures about this integral take the form of how it implicates invariants of links and 3-manifolds, and how these invariants behave in certain limits of the coupling constant k in the integral. Many conjectures of this sort can be verified through the combinatorial models. On the other hand, the really outstanding conjecture about the integral is that it exists! At the present time there is no measure theory or generalization of measure theory that supports it in full generality. Here is a formal structure of great beauty. It is also a structure whose consequences can be verified by a remarkable variety of alternative means.

As we have seen, in our earlier discussion of the Feynman path integral, it is a common situation in physics to be faced with ill-defined integrals that have strong physical meaning. In the case of Witten's function integral the relation with physics is indirect in the sense that one cannot at the present time do experiments to verify the results of formal calculations with the integral. In this way, combinatorial models for the invariants are a bit like experiments in that they provide solid (mathematially rigorous) knowledge that can be compared with the heuristics of the integral. It was for this reason that we gave a strictly mathematical description of the Vassliev invariants at the beginning of this paper. In this section we shall see that heuristic arguments with the Witten integral lead directly to Vassiliev invariants and their Lie algebraic weight systems.

In this section we will examine the formalism of Witten's approach via a generalization of our sketch of "integration without integration". In order to do this we need to consider functions $f(A)$ of gauge connections A and a notion of *functional equivalence*, $f \sim g$, taking the form $f - g = Dh$ where D is a gauge functional derivative. Since these notions need defining, we first discuss them in the context of the integrand of Witten's integral. Thus for a while, we shall speak of Witten's integral, but let it be known that this integral will soon be replaced by an equivalence class of functions just as happened in the last section. In this way, we can track the heuristics in a manner that is suitable for a mathematician.

The formalism of the Witten integral gives invariants of knots and links corresponding to each classical Lie algebra. In order to see this, we need to introduce the *Wilson loop.* The Wilson loop is an exponentiated version of integrating the gauge field along a loop K in three space that we take to be an embedding (knot) or a curve with transversal self-intersections. For this discussion, the Wilson loop will be denoted by the notation $W_K(A)$ to denote the dependence on the loop K and the field A. It is usually indicated by the symbolism $tr(Pe^{\oint_K A})$. Thus

$$W_K(A) = tr(Pe^{\oint_K A}).$$

Here the P denotes path ordered integration - we are integrating and exponentiating matrix valued functions, and so must keep track of the order of the operations. The symbol tr denotes the trace of the resulting matrix. This Wilson loop integration exists by normal means and does not require functional integration.

Understanding this approach to Witten's formulation depends upon key facts relating the curvature of the gauge field to both the Wilson loop and the Chern-Simons action. There are two ways that curvature appears in the context of the Witten integral. The variation of the Chern-Simons action with respect to the gauge field A is directly related to the curvature tensor for the gauge field. See below. This means that if the field varies, then the Lagrangian detects the curvature. The second way that curvature appears is when one evaluates a very small Wilson loop. Then the value of the loop becomes a product of the local curvature of the field, multiplied by the area spanned by the loop. The significance of this second appearance of curvature is that if we make a small deformation of the knot, then the value of the Wilson loop will change in accordance with the curvature of the gauge field near the place of deformation. *Changing the loop brings forth curvature, and changing the field brings forth curvature.* We are going to look at the mathematics of the integral and see that these two curvature sensitivities act to cancel each other out, making the integral of the Wilson loop, weighted by the exponentiated Chern-Simons action a topological invariant of the knot K. Viewing the Witten integral in terms of curvature in this way is due to Lee Smolin [**38**] (see also [**37**]).

Here is the local coordinate structure of the gauge field $A(x)$, where x is a point in three-space: We write $A(x) = A_k^a(x)T_a dx^k$ where the index a ranges from 1 to m with the Lie algebra basis $\{T_1, T_2, T_3, ..., T_m\}$. The index k goes from 1 to 3. For each choice of a and k, $A_k^a(x)$ is a smooth function defined on three-space. In $A(x)$ we sum over the values of repeated indices. The Lie algebra generators T_a are matrices corresponding to a given representation of the Lie algebra of the gauge group G. We assume some properties of these matrices as follows:

1. $[T_a, T_b] = if^{abc}T_c$ where $[x, y] = xy - yx$, and f^{abc} (the matrix of structure constants) is totally antisymmetric. There is summation over repeated indices.

2. $trace(T_aT_b) = \delta_{ab}/2$ where δ_{ab} is the Kronecker delta ($\delta_{ab} = 1$ if $a = b$ and zero otherwise).

Wilson loop mechanics. Since the crux of the stucture of the Witten integral (for knot invariants) is the Wilson loop, lets take a closer look. In thinking about the Wilson loop $W_K(A) = tr(Pe^{\oint_K A})$, it is helpful to recall Euler's formula for the exponential:

$$e^x = lim_{n\to\infty}(1 + x/n)^n.$$

In analogy to Euler's formula, we convert the Wilson loop into the trace of a product of matrices. The Wilson loop is the limit, over partitions of the loop K, of products of the matrices $(1 + A(x))$ where x runs over the partition. Thus we can write symbolically,

$$W_K(A) = tr(\prod_{x\in K}(1 + A(x)))$$
$$= tr(\prod_{x\in K}(1 + A_k^a(x)T_a dx^k)).$$

It is to be understood that a product of matrices around a closed loop connotes the trace of the product. The ordering is forced by the one dimensional nature of the loop. Insertion of a given matrix into this product at a point on the loop is then a well-defined concept. If T is a given matrix then it is understood that $TW_K(A)$ denotes the insertion of T into some point of the loop.

From this point of view, the gauge field $A(x)$ can be regarded as a *connection* governing the parallel translation of vectors $v(x) = (v, x)$ in a space $V \times R^3$ with $dim(V)$ equal to the dimension of the space on which the Lie algebra matrices T_a act. Given such a vector v, its parallel translate in the direction dx is $v + A(x)v = (1 + A(x))v$. If α is a curve in three-space, then

$$F[\alpha] = \prod_{x\in\alpha}(1 + A(x))$$

defines the operator of parallel translation along the curve α. We see that the Wilson loop is the trace of the operation of parallel translation around the knot. In differential geometry, we expect the result of parallel translation around a small loop to produce a calculation of curvature. The same thing happens in this gauge theoretic differential geometry, and we shall detail that result below. The more global parallel translation around a knotted loop contains topological information, and this information, when averaged over all the gauge fields turns an actual knot invariant. In order to see how this comes about, we nedd to look more closely at the calculus of the Wilson loops and the relationships with curvature. We begin with the following Lemma about functionally differentiating the Wilson loop.

Note that the differentiation in this Lemma is functional differentiation as described in the previous section. In Figure 13, we illustrate some of these features of the Wilson loop and make diagrammatic notations for Lie algebra elements, insertions and functional derivatives. As the reader will see shortly, these diagrammatics make it easier to follow algebraic manipulations of the formulas under Witten's integral sign. View Figure 13. The top of the figure illustrates a knot with vectors v and $(1 + A(x))v$ attached to the knot. These vectors are in the space $V \times R^3$ that is acted upon by the Lie algebra matrices in $A(x)$. This part of the figure indicates a step in the parallel translation of a vector around the loop. Below this knot diagram is the formula for the Wilson loop evaluation. Below that formula we find $T_a : V \longrightarrow V$, the action of a Lie algebra matrix as an endomorphism of the vector space V, and juxtaposed with this is our diagrammatic indication of T_a. This diagram for T_a consists in a rectangular box with input and output arrovs and a little line segment sticking out of the top of the box that is labeled with the index a. The input and output arrows correspond to the action of T_a on a vector. This box formalism for a linear transformation is exactly the same as the formalism we used for matrices in the section on Vassiliev invariants. Before reading further, the reader should look carefully at this formalism and note that it is fitting for describing matrix algebra. Finally, the figure indicates diagrams for *inserting* Lie algebra generators into the Wilson loop. Since the Wilson loop is a (limit of a) trace of a product of matrices, it is possible to choose a matrix T_a and put it in the product at a certain place x, taking then the product of all prececessors of x, then multiplying by T_a, then taking the product of all successors to x. This is exactly the process discussed in the Lemma. The notation $T_a dx^k W_K(A)$ means that $T_a dx^k$ is inserted into an approrpriate place in the ordered product for the Wilson loop. This is made explicit in the proof of the Lemma below and it indicated in the diagrams of Figure 13.

Lemma. We have the following formula for the variation of the Wilson loop with respect to the gauge field:

$$\delta W_K(A)/\delta(A_k^a(x)) = T_a dx^k W_K(A).$$

Varying the Wilson loop with respect to the gauge field results in the insertion of an infinitesimal Lie algebra element into the loop.

Proof.

$$\begin{aligned}
\delta W_K(A)/\delta(A_k^a(x)) &= \frac{\delta}{\delta(A_k^a(x))} \prod_{y \in K} (1 + A_k^a(y) T_a dy^k) \\
&= \prod_{y<x \in K} (1 + A_k^a(y) T_a dy^k)[T_a dx^k] \prod_{y>x \in K} (1 + A_k^a(y) T_a dy^k) \\
&= T_a dx^k W_K(A).
\end{aligned}$$

Curvature. The result of evaluating a Wilson loop about a very small planar circle around a point x is proportional to the area enclosed by this circle times the corresponding value of the curvature tensor of the gauge field evaluated at x. The curvature tensor is written

$$F^a_{rs}(x)T_a dx^r dy^s.$$

It is the local coordinate expression of $F = dA + A \wedge A$.

In Figure 14 we illustrate some points that are related to this assertion. View Figure 14. The first line indicates our diagram for the curvature tensor F^a_{rs} to be discussed below. Next to this diagram is our diagram for the differential dx^k as a triangle with a line segment that can carry the index k. In the next line we have the diagram for the epsilon tensor ϵ_{ijk}. The epsilon tensor takes values 1 or -1 if ijk is a permutation of 123 with the sign the sign of the permutation. If any two indices in ijk are equal, the epsilon tensor takes the value 0. The next line of Figure 14 illustrates our diagrammatic for the gauge derivative of the Wilson loop, as in the previous Lemma. This Lemma can be expressed directly as the equality in this diagram with the insertion into the Wilson loop indicated by a Lie algebra box and the extra differential carried along as the triangle diagram. The next line indicates how the Wilson loop changes when there is a small topological deformation of the underlying curve. The difference between the Wilson loop of K and the Wilson loop of K' (a slight deformation of K) is given by a Lie algebra insertion into the Wilson loop that is coupled with the curvature tensor. Finally, the last formula in Figure 14 expresses the fact (to be stated formally below) that *the gauge derivative of the Chern-Simons Action, tied into an epsilon tensor, yields the curvature tensor.* All these facts about the curvature are indicated in diagrams for ease of remembering and for the sake of seeing the pattern in calculations.

Here is the formal statement of the relationship of curvature and the Chern-Simons action: *The variation of the Chern-Simons action S with respect to the gauge potential at a given point in three-space is related to the curvature tensor at that point by the following formula:*

$$F^a_{rs}(x) = \sum_{t=1,2,3} \epsilon_{rst}\delta S/\delta(A^a_t(x)).$$

Here ϵ_{abc} is the epsilon symbol for three indices as discussed above. This formula is indicated diagrammatically at the bottom of Figure 14. See [**21**] for a derivation of this fact. It is tempting to look at the version of curvature in terms of differntial forms

$$F = dA + A \wedge A$$

and the expression for the Chern-Simons Lagrangian

$$L = A \wedge dA + (2/3)A \wedge A \wedge A$$

and note that differentiating L with respect to A is almost F. This is a hint in the direction of the exact formula, which takes more work to verify.

Variation of the Wilson loop. Consider a given Wilson line $W_K(A)$. Ask how its value will change if K is deformed infinitesimally in the neighborhood of a point x on the loop. Approximate the change via the curvature tensor as discussed above. Regard the point x as the place of curvature evaluation. Let $\delta W_K(A)$ denote the change in the value of the Wilson loop. Then $\delta W_K(A)$ is given by the insertion formula

$$\delta W_K(A) = dx^r dx^s F^{rs}_a(x)T_a W_K(A).$$

This is the first order approximation to the change in the Wilson loop. Refer again to Figure 14 for our diagram of this insertion formula.

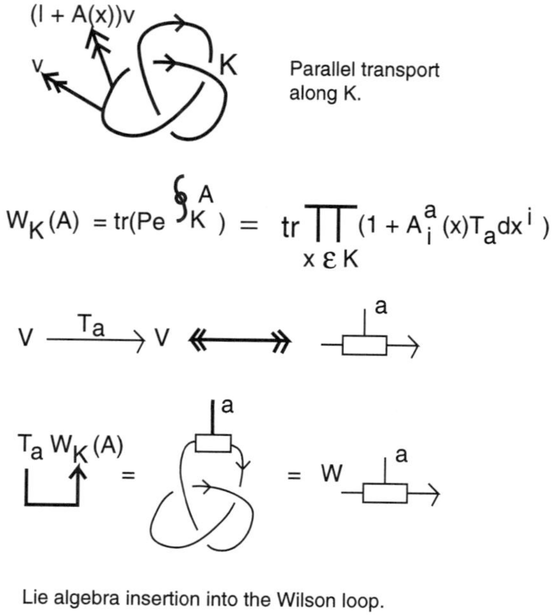

FIGURE 13. **Wilson loop and Lie algebra insertion**

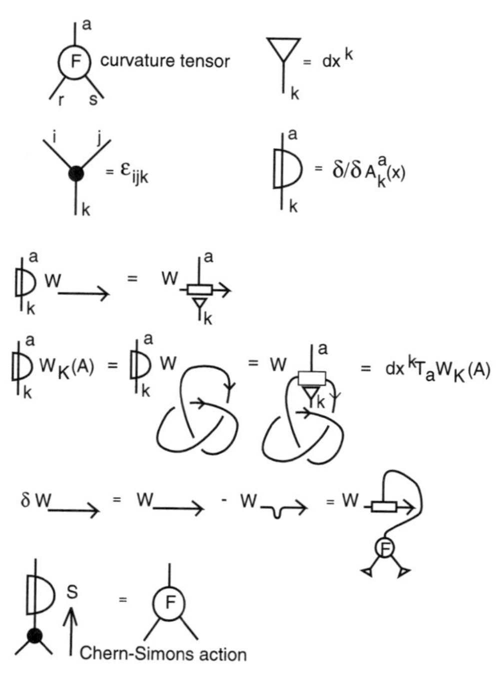

FIGURE 14. **Curvature from variation of the Chern-Simons action**

Witten's functional integral for link invariants in a 3-manifold M is expressed in terms of the Wilson loop functional.

$$Z(M,K) = \int DAe^{(ik/4\pi)S(M,A)}tr(Pe^{\oint_K A})$$
$$= \int DAe^{(ik/4\pi)S}W_K(A).$$

Here $S(M,A)$ is the Chern-Simons Action, descirbed in the previous discussion. We abbreviate $S(M,A)$ as S and write $W_K(A)$ for the Wilson loop. Unless otherwise mentioned, the manifold M will be the three-dimensional sphere S^3.

An analysis of the formalism of this functional integral reveals quite a bit about its role in knot theory. One can determine how the Witten integral behaves under a small deformation of the loop K. We delineate this in the Theorem below and in diagramamtic form in Figure 15. Before reading the Theorem, the reader should look at Figure 15. The calculation in the Theorem is diagrammed in the figure: We begin with the integration of the geometric variation of the Wilson loop. This gives a curavature insertion in the Wilson loop. The curvature tensor is re-written in terms of differentiating the Chern-Simons action S (and the use of the epsilon tensor as in Figure 14). This differentiation is rewritten as a derivative of the exponentiated action, and then we apply integration by parts, throwing the derivative back to the Wilson loop. Differentiating the Wilson loop leads to another Lie algebra insertion. The end result is an integration of a Wilson loop with two correlated Lie algebra insertions, multiplied by a volume form. This volume form will vanish if the original deformation does not involve three spatial dimensions. We conclude that for the analogues of the second and third Reidemeister moves, the functional integral is invariant. It is an invariant of regular isotopy. Framing changes will be discussed after the proof of the Theorem. Note how curvature coming from geometric variation plus curvature related to variation of the gauge field interact to cancel the local effects and produce topological invariance. This entire argument works at the level of equivalence classes of gauge functionals as desrcribed in the previous section.

Theorem.

(1) Let $Z(K) = Z(S^3, K)$ and let $\delta Z(K)$ denote the change of $Z(K)$ under an infinitesimal change in the loop K. Then

$$\delta Z(K) = (4\pi i/k) \int dAe^{(ik/4\pi)S}[Vol]T_aT_aW_K(A)$$

where $Vol = \epsilon_{rst}dx^r dx^s dx^t$.

The sum is taken over repeated indices, and the insertion is taken of the matrices T_aT_a at the chosen point x on the loop K that is regarded as the center of the deformation. The volume element $Vol = \epsilon_{rst}dx_r dx_s dx_t$ is taken with regard to the infinitesimal directions of the loop deformation from this point on the original loop.

(2) The same formula applies, with a different interpretation, to the case where x is a double point of transversal self intersection of a loop K, and the deformation consists in shifting one of the crossing segments perpendicularly to the plane of intersection so that the self-intersection point disappears. In this case, one T_a is inserted into each of the transversal

crossing segments so that $T_aT_aW_K(A)$ denotes a Wilson loop with a self intersection at x and insertions of T_a at $x+\epsilon_1$ and $x+\epsilon_2$ where ϵ_1 and ϵ_2 denote small displacements along the two arcs of K that intersect at x. In this case, the volume form is nonzero, with two directions coming from the plane of movement of one arc, and the perpendicular direction is the direction of the other arc.

Proof.

$$\begin{aligned}\delta Z(K) &= \int DAe^{(ik/4\pi)S}\delta W_K(A)\\ &= \int DAe^{(ik/4\pi)S}dx^r dy^s F^a_{rs}(x)T_aW_K(A)\\ &= \int DAe^{(ik/4\pi)S}dx^r dy^s[\epsilon_{rst}\frac{\delta S}{\delta(A^a_t(x))}]T_aW_K(A)\\ &= (-4\pi i/k)\int DA\frac{\delta}{\delta(A^a_t(x))}[e^{(ik/4\pi)S}]\epsilon_{rst}dx^r dy^s T_aW_K(A)\\ &= (4\pi i/k)\int DAe^{(ik/4\pi)S}\epsilon_{rst}dx^r dy^s\frac{\delta}{\delta(A^a_t(x))}[T_aW_K(A)]\end{aligned}$$

(integration by parts and the boundary terms vanish)

$$= (4\pi i/k)\int DAe^{(ik/4\pi)S}[Vol]T_aT_aW_K(A).$$

This completes the formalism of the proof. In the case of part 2., a change of interpretation occurs at the point in the argument when the Wilson line is differentiated. Differentiating a self intersecting Wilson line at a point of self intersection is equivalent to differentiating the corresponding product of matrices with respect to a variable that occurs at two points in the product (corresponding to the two places where the loop passes through the point). One of these derivatives gives rise to a term with volume form equal to zero, the other term is the one that is described in part 2. This completes the proof of the Theorem.

Remark. We have shown that the result of a topological variation has an analytic expression that is zero if the topological variation does not create a local volume. Thus we have shown that the intergral of $e^{(ik/4\pi)S(A)}W_K(A)$ is topologically invariant as long as the curve K is moved by the local equivalent of regular isotopy.

In the case of switching a crossing the key point is to write the crossing switch as a composition of first moving a segment to obtain a transversal intersection of the diagram with itself, and then to continue the motion to complete the switch. Up to the choice of our conventions for constants, the switching formula is, as shown below (See Figure 16).

$$\begin{aligned}Z(K_+) - Z(K_-) &= (4\pi i/k)\int DAe^{(ik/4\pi)S}T_aT_a < K_{**}|A >\\ &= (4\pi i/k)Z(T^aT^aK_{**}),\end{aligned}$$

where K_{**} denotes the result of replacing the crossing by a self-touching crossing. We distinguish this from adding a graphical node at this crossing by using the

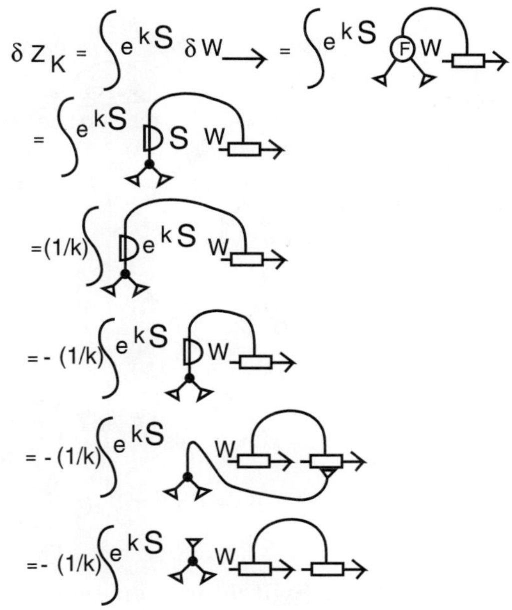

FIGURE 15. **Variation of the integral with respect to a topological deformation**

double star notation. A key point is to notice that *the Lie algebra insertion for this difference is exactly what is done (in chord diagrams) to make the weight systems for Vassiliev invariants* (without the framing compensation). Thus the formalism of the Witten functional integral takes one directly to these weight systems in the case of the classical Lie algebras. In this way the functional integral is central to the structure of the Vassiliev invariants.

To see the relationship with Vassiliev invariants, let $K_\bullet$ denote the result of replacing the crossing in K_+ or K_- by a 4-valent node. Make the following definition

$$Z(K_\bullet) = Z(T^a T^a K_{**})$$

as in Figure 16. We then have the difference formula

$$Z(K_+) - Z(K_-) = (4\pi i/k) Z(K_\bullet).$$

If $Z(K)$ is expressed as a series in powers of $1/k$, the coefficients will be Vassiliev invariants of finite type by the same arguments that we used in the section on Vassiliev invariants (divisibility of $Z(G)$ by $(1/k)^n$ if the graph G has n nodes). The fact that the Lie algebra insertion is built into the definition of $Z(K_\bullet)$, makes it possible to see that the weight systems for these invariants are exactly the same as the weight systems associated with Lie algebras that we described in Section 3.

We started this journey in Section 3 by showing how one could in principle construct link invariants in relation to Lie algebras via the translation of the Reidemeister moves for graph embeddings into the four-term relations for the weight systems. These four-term relations are directly related to commutator closure and to Jacobi identity for Lie algebras. We now meet this construction from the other end, seeing the Lie algebra weight systems at the base of the functional integral

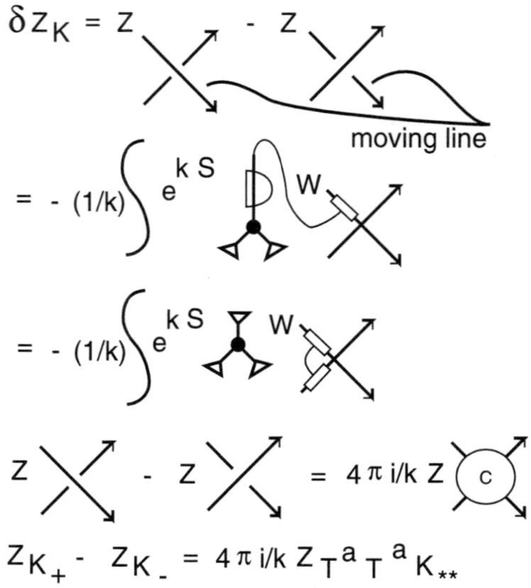

FIGURE 16. **Switching a crossing**

approach due to Witten. His functional integral then supplies a way to integrate the weight systems to produce link invariants. We are going to stop at this point, but the reader should know that the story does not stop! The Witten integral is only a half-way point because it is a hueristic formalism. One can carry the heuristics further to the Feynman diagram expansion of the Witten integral [**23**] and uncover the well-known Kontsevich integrals [**4**] that were used by Kontsevich to give a rigorous basis for the Vassiliev invariants (in terms of given weight systems). The Kontsevich integrals can be developed on their own right and the heuristic scaffolding can be removed. The purpose of this paper has been to show how the knot invariants are related to Lie algebras, physics and the heuristics of functional integration.

Remark on integration without integration. Note that in terms of functional equivalence (integration without integration as discussed above and in the previous section), we have shown that

$$e^{(ik/4\pi)S(A)}\delta W_K(A) \sim e^{(ik/4\pi)S(A)}[Vol]T_aT_aW_K(A).$$

In terms of functional equivalence, we have shown that the equivalence class of a topological variation has an analytic expression that is zero if the topological variation does not create a local volume. *Thus we have shown that the functional equivalence class of* $e^{(ik/4\pi)S(A)}W_K(A)$ *is topologically invariant as long as the curve* K *is moved by the local equivalent of regular isotopy.* This is the sort of conclusion that we get *without* needing the existence of the functional integral.

7. The Loop Transform and Quantum Gravity

In [**3**] and other publications by Ashtekar, Rovelli, Smolin and their colleagues, the loop transform is used to study a reformulation and quantization of Einstein gravity. The differential geometric gravity theory of Einstein is reformulated in terms of a background gauge connection A. In the quantization, the Hilbert space of quantum wave functions consists in complex valued functions $\psi(A)$ that are required to satisfy the constraints

$$G\psi = 0$$

and

$$H\psi = 0$$

for a pair of operators G and H called respectively, the *diffeomorhism constraint* and the *Hamiltonian constraint.* We shall exhibit these operators below. There is not room here to detail how general relativity is reformulated in terms of gauge theory, but we can indicate how functional integration like the Witten functional integral of the last section comes into play in this theory of *loop quantum gravity.* Surveys of loop quantum gravity that are in the spirit of this section can be found at [**33–35**].

Suppose that $\psi(A)$ is a (complex valued) function defined on gauge fields. Then we define formally the *loop transform* $\widehat{\psi}(K)$, a function on embedded loops in three-dimensional space, by the formula

$$\widehat{\psi}(K) = \int DA\psi(A)W_K(A).$$

In other words, we take $\widehat{\psi}(K)$ to be the functional equivalence class of $\psi(A)W_K(A)$ in the sense of the previous section.

$$\widehat{\psi}(K) \sim \psi(A)W_K(A).$$

If Δ is a differential operator defined on $\psi(A)$, then we can use this integral transform to shift the effect of Δ to an operator on loops via integration by parts:

$$\begin{aligned}\widehat{\Delta\psi}(K) &= \int DA\Delta\psi(A)W_K(A)\\ &= -\int DA\psi(A)\Delta W_K(A).\end{aligned}$$

Again, this is a statement about the functional equivalence classes:

$$\widehat{\Delta\psi}(K) \sim (\Delta\psi(A))W_K(A) \sim -\psi(A)(\Delta W_K(A)).$$

When Δ is applied to the Wilson loop the result can be an understandable geometric or topological operation. One can illustrate this situation with operators G and H.

$$\begin{aligned}G &= -F^a_{ij}dx^i\delta/\delta A^a_j(x),\\ H &= -\epsilon_{ars}F^a_{ij}\delta/\delta A^s_i\delta/\delta A^r_j\end{aligned}$$

with summation over the repeated indices. Each of these operators has the property that its action on the Wilson loop has a geometric or topological interpretation. Here we show in Figures 17 and 18 what happens with the diffeomorphism constraint G. Viewing Figure 18 we see the diagrammatic verision of G undergoing the loop transform. Note that after the integration by parts, G acts on the Wilson loop and, after differntiation, produces a Lie algebra insertion linked with the curvature tensor that we have previously identified as the result of a geometric variation of the Wilson loop! We have the formula

$$\widehat{G\psi}(K) = \delta\widehat{\psi}(K)$$

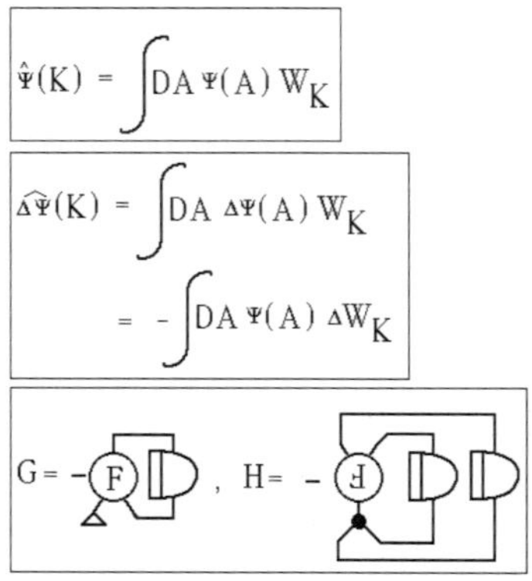

FIGURE 17. **The loop transform and the operators G and H**

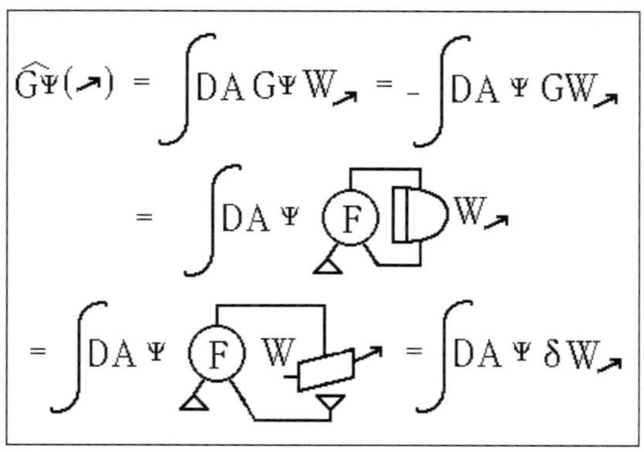

FIGURE 18. **The loop transform of g is a topological variation**

where this variation refers to the effect of varying the loop K by a small deformation. As we saw in the previous section, this means that if $\widehat{\psi}(K)$ is a topological invariant of knots and links, then $\widehat{G\psi}(K) = 0$ for all embedded loops K. This condition is a transform analogue of the equation $G\psi(A) = 0$. This equation is the differential analogue of an invariant of knots and links. It may happen that $\delta\widehat{\psi}(K)$ is not strictly zero, as in the case of our framed knot invariants. For example with

$$\psi(A) = e^{(ik/4\pi)\int tr(A\wedge dA + (2/3)A\wedge A\wedge A)}$$

we conclude that $\widehat{G\psi}(K)$ is zero for flat deformations (in the sense of the previous section) of the loop K, but can be non-zero in the presence of a twist or curl. In this sense the loop transform provides a subtle variation on the strict condition $G\psi(A) = 0$.

Nevertheless, it turns out that the exponentated Chern-Simons action, $\psi(A)$ above, is a significant quantum state for loop quantum gravity and the theory is linked in a remarkable way with both invariants of knots and links, and with the Chern-Simons-Witten approach to these invariants.

8. Epilogue

This paper has been an introduction to the relationships between knots and physics related to invariants of knots and links such as the Jones polynonmial and the Vassiliev invariants. Much has been left out of the story. The purpose of this epilogue is to point to a few of the things that are left out, and give a guide to reading further.

The ideas and definition of topological quantum field theory are well expressed in the book [**2**] by Michael Atiyah and the paper [**41**] by Edward Witten. The simplest case of this idea is C. N. Yang's original interpretation of the Yang-Baxter equation. Yang articulated a quantum field theory in one dimension of space and one dimension of time in which the R-matrix giving the scattering ampitudes for an interaction of two particles whose (let us say) spins corresponded to the matrix indices so that R_{ab}^{cd} is the amplitude for particles of spin a and spin b to interact and produce particles of spin c and d. Since these interactions are between particles in a line, one takes the convention that the particle with spin a is to the left of the particle with spin b, and the particle with spin c is to the left of the particle with spin d. If one follows the concatenation of such interactions, then there is an underlying permutation that is obtained by following strands from the bottom to the top of the diagram (thinking of time as moving up the page). Yang designed the Yang-Baxter equation for R so that the amplitudes for a composite process depend only on the underlying permutation corresponding to the process and not on the individual sequences of interactions. In Yang's picture, state spaces are associated with points (zero-dimensional manifolds) and linear transformations of states to braids, which can be regarded as *cobordisms of zero-dimensional manifolds* (the braid is one-dimensional and its boundary consists of the initial and final collections of points). Just so, one can associate state spaces to n-manifolds and linear transformations to cobordisms of n-manifolds. This is the basic idea for constructing topological quantum field theories.

Link invariants can be constructed by using solutions to the Yang-Baxter equation to supply vertex weights for state summations that generalize the bracket polynomial. This is a significant conduit between statistical mechanics and knot theory. See [**18, 21, 31, 36, 39**].

It turns out that the algebraic properties of a topological quantum field theory give it enough power to rigourously model three manifold invariants described by the Witten integral. This is done by regarding the three-manifold as a union of two handlebodies with boundary an orientable surface S_g of genus g. The surface is divided up into trinions. A *trinion* is a surface with boundary that is topologically equivalent to a sphere with three punctures. It turns out that there is a way to associate a vector space $V(S_g)$ to a surface with a trinion decomposition, such that the isomorphism class of the vector space $V(S_g)$ does not depend upon the choice of decomposition. Furthermore, if a closed three-manifold M^3 is decomposed along a surface S_g into the union of M_- and M_+ where these parts are otherwise disjoint three-manifolds with boundary S_g, then one obtains vectors $|M_+\rangle$ and $|M_-\rangle$ in $V(S_g)$ such that the inner product $I(M) = \langle M_-|M_+\rangle$ is, up to normalization, an invariant of the three-manifold M^3. Knots and links can be incorporated as well, so that one obtains a source of invariants $I(M^3, K)$ of knots and links in orientable three-manifolds. The invariant $I(M^3, K)$ can be formally compared with the Witten integral

$$Z(M^3, K) = \int DAe^{(ik/4\pi)S(M,A)}W_K(A).$$

It can be shown that up to limits of the heuristics, $Z(M, K)$ and $I(M^3, K)$ are essentially equivalent for appropriate choice of gauge groups.

One can imagine a physics where the geometrical space is two dimensional and the braiding of particles corresponds to their interactions through circulating around one another in the plane. One expects that physical situations involving $2 + 1$ spacetime will be related to braids where the braid is the world line of a collection of particles moving around one another in the plane. It is thought for example, that aspects of the quantum Hall effect will be related to topological quantum field theory [**30, 40**]. *Anyons* are particles that do not just change their wave functions by a sign under interchange, but rather by a complex phase or even a linear combination of states. It is hoped that topological models of this kind will describe applicable physics, in particular for quantum computation [**16, 24–26**]. It may be that elementary particles are directly related to the properties of braids [**5**].

People often ask whether knot theory is related to string theory. String theory is based on the idea that one could replace a point particle by a string, where this string could be imagined as a tiny circle whose vibrational modes give rise to all the particle properties. Why shouldn't this circle be knotted, so that the strings have topological states and the vibrational modes of the string depend upon the topology of the embedding of the knot? This is a very enticing idea, but it has not been articulated in practice for the following reasons: The strings (for technical reasons) are embedded in a ten dimensional space with four dimensions of space-time and six internal dimensions related to the quantum field theory of the string. Unfortunately circles embedded in a space of dimension greater than three are unknotted. So the knotting of a string would be invisible in such a model. Of course if you imagined that the string was really in three space,knotted and that the knotting was correlated in some way with the way the extra dimensions are wrapped up in micro-space, then maybe you could make a theory of knotted strings. Perhaps the reader will find a way. There are very interesting technical relationships between string theory and knot invariants. For this, the reader should consult the paper [**28**] which reviews the relation between Chern-Simons gauge theory and topological string theory on noncompact Calabi-Yau spaces.

Finally, we cannot resist mentioning a series of papers about the possibility of knotted glueballs. A *glueball* is a loop of gluon flux, the field that binds quarks. It is theoretically possible for a quark string (quark annd antiquark bound by a string of gluon flux) to interact with itself so that the quark ends annihilate one another, and one is left with a closed loop of gluon flux. This is a glueball. Why can't the quark-antiquark string get knotted before it self-interacts? Then the result of the interaction would be a loop of knotted gluon flux. Read the following papers [**7–9**].

References

[1] J.W. Alexander. Topological invariants of knots and links. Trans. Amer. Math. Soc. 20 (1923), 275-306.
[2] M.F. Atiyah, *The Geometry and Physics of Knots*, Cambridge University Press, 1990.
[3] Ashtekar, Abhay, Rovelli, Carlo and Smolin, Lee [1992], "Weaving a Classical Geometry with Quantum Threads", *Phys. Rev. Lett.*, vol. 69, 237.
[4] D. Bar-Natan, On the Vassiliev knot invariants, *Topology* **34** (1995), 423-472.
[5] S. Bilson-Thompson, A topological model of composite preons, arXiv:hep-ph/0503213.
[6] J. Birman and X.S. Lin, Knot polynomials and Vassiliev's invariants. *Invent. Math.* 111 (1993), no. 2, 225–270.
[7] Roman V. Buniy, Thomas W. Kephart, Glueballs and the universal energy spectrum of tight knots and links, arXiv:hep-ph/0408027, *Int. J. Mod. Phys.* **A20** (2005), 1252-1259.
[8] Roman V. Buniy, Thomas W. Kephart, Universal energy spectrum of tight knots and links in physics, arXiv:hep-ph/0408025.
[9] A. Niemi, Are Glueballs Knotted Closed Strings?, arXiv:hep-th/0312133, Talk given at International Symposium on Color Confinement and Hadrons in Quantum Chromodynamics - Confinement 2003, Wako, Japan, 21-24 Jul 2003.
[10] L. Crane, 2-d physics and 3-d topology, *Comm. Math. Phys.* **135** (1991), no. 3, 615-640.
[11] R.P. Feynman, The Principle of Least Action in Quantum Mechanics, Princeton Thesis (1942).
[12] R. P. Feynman, Space-time approach to non-relativistic quantum mechanics, *Rev. Mod. Phys.*, Vol. 20 (1948), 367-387.
[13] R. P. Feynman and A. R. Hibbs, Quantum Mechanics and Path Integrals, McGraw Hill Pub. Co., Massachusetts (1965).
[14] R. P. Feynman, QED – The Strange Theory of Light and Matter, Princeton University Press (1985).
[15] R. P. Feynman, Feynman's Thesis - A New Approach to Quantum Theory, edited by L. Brown, World Scientific Pub. Co. (reprint of the 1942 Thesis).
[16] M. H. Freedman, A. Kitaev, Z. Wang, Simulation of topological field theories by quantum computers, *Commun. Math. Phys.*, **227**, (2002), 587-60, quant-ph/0001071.
[17] V.F.R. Jones, A polynomial invariant for links via von Neumann algebras, Bull. Amer. Math. Soc. **129** (1985), 103–112.
[18] V.F.R. Jones, On knot invariants related to some statistical models, *Pacific J. Math.*, bf 137 (1989), 311-336.
[19] L.H. Kauffman, State models and the Jones polynomial, Topology **26** (1987), 395–407.
[20] L.H. Kauffman, *Temperley-Lieb Recoupling Theory and Invariants of Three-Manifolds*, Princeton University Press, Annals Studies **114** (1994).
[21] L.H. Kauffman, *Knots and Physics*, World Scientific Publishers (1991), Second Edition (1993), Third Edition (2002).
[22] L.H. Kauffman, Functional Integration and the theory of knots, J. Math. Physics, Vol. 36 (5), May 1995, 2402 - 2429.
[23] L.H. Kauffman, Functional Integration, Kontsevich Integral and Formal Integration. J. Korean Math. Soc. 38 (2001), No. 2, 437-468.
[24] L.H. Kauffman, Quantum computation and the Jones polynomial, in *Quantum Computation and Information*, S. Lomonaco, Jr. (ed.), AMS CONM/305, 2002, 101–137.
[25] L.H. Kauffman and S. J. Lomonaco Jr., Quantum entanglement and topological entanglement, New Journal of Physics **4** (2002), 73.1–73.18 (http://www.njp.org/).
[26] L. H. Kauffman and S. J. Lomonaco Jr., q-Deformed Spin Networks, knot polynomials and topological quantum computation *JKTR* Vol. 16, No. 3 (March 2007), 267-332.
[27] Louis H. Kauffman, Knot theory and physics. article for Encyclopedia of Physical Science and Technology, Third Edition, Volume 8, (2002), 199-245.
[28] M. Marino, Chern-Simons Theory and Topological Strings, hep-th/0406005, *Rev.Mod.Phys.* **77** (2005), 675-720.
[29] A. Marzuoli and M. Rasetti, Spin network quantum simulator, *Physics Letters A* **306** (2002), 79–87.
[30] G. Moore and N. Read, Nonabelions in the fractional quantum Hall effect, *Nucl. Phys. B360* (1991), 362-396.

[31] T. Ohtsuki, "Quantum Invariants", World Scientific Pub. Co. (2002).
[32] R. Penrose, Angular momentum: An approach to Combinatorial Spacetime, In *Quantum Theory and Beyond*, edited by T. Bastin, Cambridge University Press (1969).
[33] J. Pullin, Knot theory and quantum gravity in loop space: a primer. arXiv:hep-th/930128.
[34] R. Gambini and J. Pullin, "Loops, Knots, Gauge Theories and Quantum Gravity", Cambridge University Press (1996).
[35] C. Rovelli, "Quantum Gravity", Cambridge University Press (2004).
[36] N.Y. Reshetikhin and V. Turaev. Invariants of Three Manifolds via link polynomials and quantum groups. Invent. Math. 103, 547-597 (1991).
[37] P. Cotta-Ramusino, E. Guadagnini, M. Martellini, M. Mintchev, Quantum field theory and link invariants, *Nucl. Phys. B* **330**, Nos. 2-3 (1990), 557-574.
[38] L. Smolin, Link polynomials and critical points of the Chern-Simons path integrals, *Mod. Phys. Lett. A*, Vol. 4,No. 12, (1989), 1091-1112.
[39] V. G. Turaev, The Yang-Baxter equation and invariants of links, *Invent. Math.* **92** (1988), 527-553.
[40] F. Wilczek, *Fractional Statistics and Anyon Superconductivity,* World Scientific Publishing Company (1990).
[41] E. Witten, Quantum field Theory and the Jones Polynomial, *Commun. Math. Phys.*, vol. 121, (1989), 351-399.

Department of Mathematics, University of Illinois at Chicago, 851 South Morgan Street, Chicago, Illinois 60607-7045

Current address: Department of Mathematics, University of Illinois at Chicago, 851 South Morgan Street, Chicago, Illinois 60607-7045

E-mail address: `kauffman@uic.edu`

Proceedings of Symposia in Applied Mathematics
Volume **66**, 2009

Synthetic Single-Stranded DNA topology

Nadrian C. Seeman

Abstract. The double helical structure of DNA makes it an excellent component for the molecular construction of topological targets, such as knots and catenanes. This capability is facilitated by the ability to design and synthesize DNA molecules that contain branched helix axes, rather than the linear molecules that are the basic repository of genetic information in living systems. DNA has been used to synthesize polyhedral catenanes, a variety of single-stranded knots, and Borromean rings. A protocol is described for generating DNA motifs that uses the biological process of reciprocal exchange between DNA molecules. Using this protocol, DNA strands can be combined to construct a variety of robust motifs that can, in turn, self-assemble into two-dimensional, and three-dimensional crystalline lattices. In addition to static structures, these motifs can also be used as components of DNA-based nanomechanical devices.

1. Introduction

We are all familiar with DNA, the carrier of genetic information in living organisms. From an obscure understanding in the first half of the twentieth century, its double helical structure has emerged as a ubiquitous icon of contemporary civilization, similar to the pyramids of Egypt or pre-Columbian America, the cathedrals of medieval Europe, the Taj Mahal or the Great Wall of China. Many people are aware that genetic information is linearly encrypted inside the helix, *via* the order of base pairs: School children today are taught that there are four bases, adenine (A), guanine (G), cytosine (C) and thymine (T). These bases occur within the double helix as complementary hydrogen bonded base pairs of the same size; A is always paired with T and G is always paired with C.

It is less well appreciated that DNA is perhaps the most versatile and convenient polymer from which to design and synthesize topological target species. In this chapter, we will describe two different types of DNA topology, branching and linking. We will describe a protocol that enables us to design a variety of DNA

2000 *Mathematics Subject Classification.* Primary 92C40.

This research has been supported by grants GM-29554 from NIGMS, grants DMI-0210844, EIA-0086015, CCF-0432009, CCF-0523290 and CTS-0548774, CTS-0608889, CCF-0726378 from the NSF, 48681-EL and MURI W911NF-07-1-0439 from ARO, DE-FG02-06ER64281 from DOE (Subcontract from the Research Foundation of SUNY), and a grant from the W.M. Keck Foundation.

molecules that contain crossovers between helices. There are specific procedures that will be discussed enabling us to engineer the sequences of DNA molecules, and thereby to direct the formation of these species, and of fused branch species, such as stick polyhedra, as well as linked species, such as knots, catenanes (known in some parts of the literature as links) and Borromean rings. By following simple procedures that mix the drawing board (for concepts), the computer (for graphics and sequence assignment) and the wet laboratory, it is possible to design complex DNA motifs that can be used in synthesizing physical DNA structures and topologies. Two-dimensional arrays of branched DNA can include molecular chain mail or interwoven species as substructures. Single-stranded DNA topology is intimately related to the action of robust DNA-based nanomechanical devices.

2. DNA helicity leads to linked topological species

The DNA double helix consists of two strands of opposite polarity, wrapped around each other [**1**]. One end of each strand is called the 5′ end, and the other end is called the 3′ end; they are different, thereby leading to the polarity. The individual nucleic acid repeating unit consists of a base attached to a sugar-phosphate, which is called a nucleotide. The helix is about 20 Å wide, the vertical separation of the base pairs is about 3.4 Å, and the periodicity is 10-10.5 nucleotides/turn; for simplicity, we will usually use 10 nucleotides/turn here. The double helical nature of the structure has a subtle implication: The two molecules wrap around each other so as to create topological features involving the individual strands. For example, a 1000-base pair piece of DNA whose ends are joined to produce a circular molecule that consists of two strands that are linked 100 times.

The fundamental units of any topological construction are unit tangles [**2**]. In a two-dimensional projection, each of these unit tangles appears as a crossing or a node, where a strand intersects itself or another strand; the strand closer to the viewer is usually drawn to be continuous, and the one further away is drawn interrupted. The number, signs and relative placement of each of these nodes characterizes any particular topological construction. To achieve the construction of topological targets in chemical systems, it is necessary to construct the chemical equivalent of a unit tangle. For example, Sauvage and his colleagues have used multinuclear coordination compounds to produce unit tangles [**3**], Stoddart and his colleagues [**4, 5**] have used ionic charge-transfer complexes and Vögtle *et al.* have used hydrogen bonding combined with π-π interactions [**6**]. Figure 1a illustrates the point that a double helical half-turn of DNA is a motif that also corresponds to a unit tangle. The unit tangle is formed by two single strands, so the topology we are discussing here is single-stranded topology, distinct from double-stranded topology; the reader is referred to Cozzarelli and Wang [**7**] or Bates and Maxwell [**8**] for an introduction to that topic. The top of Figure 1a shows that there are two types of crossings, the negative node and the positive node, which are mirror images of each other. The form of DNA seen in solution and within the cell is called "B-DNA", and it is known to be right-handed [**9**]. A schematic picture of B-DNA is shown below the negative node, with arrowheads drawn on the antiparallel strands to indicate their polarities. It is clear from this drawing that the nodes generated by right-handed B-DNA are negative nodes.

DNA is capable of adopting another structure, Z-DNA, which is left-handed. Both B-DNA and Z-DNA contain sugars in the D-configuration, so Z-DNA is not the geometrical mirror image of B-DNA. However, for topological purposes, Z-DNA can be treated as the mirror image of B-DNA, because its left-handed helix produces positive nodes. Z-DNA is a very unusual structure, because its geometrical repeat unit is the dinucleotide, not the mononucleotide, leading to a zigzag backbone structure. Furthermore, only special sequences form Z-DNA readily, and even these sequences require solution conditions that promote the B $\rightarrow$ Z transition [**10**]. Thus, it is possible to target specific regions of a DNA molecule to produce left-handed helices, and to trigger them to do so by controlling solution conditions. This capability has been used to create a DNA-based nanomechanical device, described in Section 13 [**11**]. The disadvantage of using Z-DNA to produce left-handed DNA is that the number of sequences that reliably form Z-DNA is somewhat limited. Later, we will describe exciting new developments involving left-handed DNA that overcome this limitation.

3. DNA branching topology

There is another feature of DNA topology that contributes to its versatility as a topological component. This is branching topology. The DNA double helix that is found within living cells is a linear molecule, in the sense that its helix axis is unbranched. However, there are ephemeral branched structures that arise as intermediates in the processes of DNA recombination, replication and repair; probably the best-known of these is the recombination intermediate known as the Holliday junction [**12**]. The Holliday junction and other cellular branched structures are inherently unstable, because their branch points are flanked by twofold (homologous) sequence symmetry, so that the positions of the branch points can relocate through an isomerization known as branch migration [**13**]. Through the use of specially designed DNA sequences, it is possible to create stable synthetic DNA branched molecules, called "branched junctions" whose branch points are fixed [**14, 15**]. An example of a small DNA molecule containing a stable branch is illustrated in Figure 1b. These molecules have been used extensively to characterize the properties of Holliday junctions [**16, 17**].

Although the naturally-occurring Holliday junction contains four arms, there is in fact no limitation on the number of arms that a DNA branched junction can contain. Junctions have been made containing three [**18**], four [**19**], five and six [**20**], eight and twelve [**21**] arms. Both theoretical considerations [**22**] and experimental results [**21**] suggest that there is no apparent limit to this number. Nevertheless, increasing the number of double helical arms tends to require longer arms, so that eight nucleotide pairs per arm are adequate for 3-arm and 4-arm junctions, but 5-arm, 6-arm and 8-arm junctions needed longer arms (16 nucleotide pairs could be used for all three) and the 12-arm junction needed still longer arms (24 nucleotide pairs were successful). These highly branched junctions are shown in Figure 1c.

Key among their roles outside biology, branched molecules are the building blocks of structural DNA nanotechnology, the attempt to control the structure of matter on the nanometer scale through directed intermolecular interactions. Biotechnologists have used single-stranded overhangs to control the associations of linear duplex DNA molecules since the early 1970s [**23**]; this type of association is illustrated in Figure 1d, which indicates that the associations can be limited to hydrogen bonding or can be annealed to covalent bonds by (usually enzymatic) ligation. Combining branched molecules with sticky-end directed association [**14**] can lead to the assembly of N-connected stick figures and networks [**24, 25**], as shown in Figure 2a. N-connected means that N edges terminate at the vertices. The edges of these stick figures and networks consist of double-stranded DNA, and the vertices correspond to the branch points of the junctions. The connectivity of any DNA object corresponds to the number of double helical arms flanking the branch points; thus, to make a stick version of cubic close packing, one would need the 12-arm junctions recently prepared [**21**].

The selection of sequences for DNA molecules is a key element in all nucleic acid design. Any DNA molecules that are not double helical correspond to some sort of excited state, a molecule whose components are not all in their most favorable arrangements. Consequently, at least a portion of what is generally understood about the thermodynamic behavior of DNA as a molecule is being thrown out when such molecules are designed. The central notion when designing a branched molecule is to make sure that the excited state that results in the laboratory is the one targeted, not another one that turns out to be more favorable. The way in which such molecules are usually designed is by breaking up each strand into a series of overlapping shorter segments. For example, consider the branched junction in Figure 1b, where four 16-nucleotide strands assemble to form a 4-arm junction with eight nucleotide pairs per arm. Each 16-mer strand could be divided into 13 overlapping tetramers. Sequence assignment proceeds by insisting that each tetramer be unique. In addition, no tetramer that complements a tetramer flanking a branch point is permitted in the structure, thereby preventing the branch point from having the option of forming simple double helical DNA. The competition for the target structure (four octamer arms) then can come only from trimers. For a structure of this size, tetramers provide a "vocabulary" of 256 sequences (less 16 self-complementary sequences) from which to choose. Larger structures require a larger vocabulary, and hence larger sub-segments. For 4-arm junctions, it is also key to avoid the twofold sequence symmetry around the junction, so that the junction cannot undergo branch migration. This approach [**14, 15**] has been quite successful over the last quarter century, but it is worth noting that it uses strictly negative criteria, excluding certain sequences from the final strands. Approaches that entail thermodynamic minima ([**26**] e.g.) are being developed currently, but they suffer from a lack of knowledge about the thermodynamics of alternative structures. As will be seen below, the approach described above completely ignores the phenomenon of base-stacking (the tendency of base pairs to stack parallel to each other in aqueous solution) and only considers Watson-Crick base pairing. Every base can pair with every other base, including itself [**27**], and it is only in the double helical context that Watson-Crick base pairing dominates. In addition, base-stacking is ignored at one's peril. The descriptions below of antijunction and switchback molecules forming structures other than those designed are examples of a system maximizing base stacking by isomerizing, thereby forming structures other than the target structure.

The issue of single-stranded topology arises even in a simple construction, such as the one shown in Figure 2a. That drawing is a naïve schematic, in which the helicity of DNA is ignored. Figures 2b and 2c show two different solutions to the problem of incorporating helicity into this design in two dimensions. Figure 2b illustrates this construction when the vertices are separated by an exact number of double helical full-turns, or an even number of half-turns. The underlying strand structure consists of a series of linked circles, much like molecular chain mail. The linkage between each pair of circles is a function of the number of turns in the edge between them; there are two turns in each edge in this diagram, so adjacent circles are linked twice. By contrast, Figure 2c shows the consequences of separating the vertices by an odd number of half-turns. Instead of the linked circles, we see that the strand structure consists of an interwoven meshwork of strands. This example emphasizes the key point made earlier, namely that the double helical half-turn can be regarded for many purposes as the quantum of single-stranded topology, because it corresponds to a unit tangle.

The simple notions discussed in the earlier parts of this section may be appealing, but the simple branched motifs discussed here are not physically robust. The DNA molecules that flank a junction are relatively stiff, having persistence lengths of about 50 nm under standard conditions [**28**]. For a molecule that is only 2 nm wide, this is quite rigid. However, the branched junctions themselves are quite floppy [**18, 29**]; the dynamic features of a branched junction are not badly modeled by a marshmallow impaled by uncooked double helical pasta. As a consequence, the construction of species that can be built using them can be controlled topologically, but not geometrically. Examples are polyhedral catenanes with the connectivities of a cube (Figure 2d) and of a truncated octahedron (Figure 2e), which are discussed in Section 6. So as to gain geometrical control, it is necessary to generalize the motif generation process, so that species with less flexibility can be designed and constructed.

4. A motif-generation protocol

It is easy to imagine a DNA molecule with antiparallel strands to be analogous to a road with two lanes pointing in opposite directions. Imagining two such roads and having them intersect is a simple way to visualize the behavior of the single strands in a branched junction. Each lane can turn right into the intersection (in North America) and continue onwards. One can imagine more roads approaching the junction, thus leading to the highly branched molecules discussed above. Indeed, this was the approach used for the first period of structural DNA nanotechnology.

However, this approach doesn't really lead to physically useful generalization. As noted above, the reason is that individual branched junctions are not at all rigid species. The 4-arm junction has a fairly well-defined structure in the presence of magnesium, but the other junctions seem to have a floppy character around their branch points [**21**]. Although it is possible to build deltahedra from them [**30**]–[**32**], it is hard to generate robust tiles that lead to periodic or aperiodic lattices.

A more valuable approach is to use the process of reciprocal exchange, found in genetic recombination, to perform operations on adjacent helices [**33**]. Reciprocal exchange is illustrated in Figure 3a. A red strand and a blue strand are shown on the left of this drawing. Following reciprocal exchange, we have a red-blue strand and a blue-red strand on the right. Figure 3b shows that connections between motifs in the Kolams found in Hindu temples are very closely related to this same approach. The Kolams do not usually have well-defined helix axes to connect, but as will be noted later, it is sometimes possible to change the direction of the helix axis in designs. Figure 3c extends the notion of reciprocal exchange to the double helical context. Performing this operation converts two double helical molecules into a single 4-stranded molecule. The two strands where the operation has been performed are colored green in Figure 3c. It is worth noting that for right-handed B-DNA, the right-handed (negative) node is the one that forms [**34, 35**].

It is very important to realize that DNA chains are antiparallel, as indicated by the arrowheads in Figure 3c. Consequently, one can perform reciprocal exchange between strands of the same polarity or chains of opposite polarity. This is not important when there is only a single crossover. However, if one is to perform more than a single reciprocal exchange, then different structures and topologies will result, depending on whether the exchange takes place between strands of the same polarity or strands of opposite polarity. This is illustrated in Figures 3d and 3f. Here, we perform two reciprocal exchanges between the molecules, to produce what are known as DX molecules. In Figure 3d, reciprocal exchange is performed between strands of the same polarity and in Figure 3f, reciprocal exchange is performed between strands of opposite polarity. Note that in the exchange between strands of the same polarity, four strands are produced, but five strands result when the exchange occurs between strands of the opposite polarity, as seen with the blue strand in Figure 3f. This diagram has not taken all of the possibilities into account, and in fact there are five possible isomers of DX molecules that can result when two reciprocal exchanges occur between two DNA double helices, three involving strands of the same polarity, and two involving strands of opposite polarity [**36**]. As a practical matter, when the crossovers are close together, the DX molecules that result from opposite polarity crossovers are much better behaved in the laboratory than those that result from identical polarity crossovers; hence there has been very little experimental work done with identical polarity DX molecules, although they are known to be meiotic intermediates [**37**] when the crossovers are separated by large distances. Applications of DX molecules are discussed below.

Figure 3e illustrates the results when reciprocal exchange occurs between every possible point of juxtaposition, in the case of same-polarity crossover. This leads to the remarkable PX structure, which looks like two double helices that have been wrapped around each other. Thus, there appears to be a red double helix wrapped around a green double helix. This is not exactly what happens. As one reads from top to bottom in either helix in the PX molecule, there is a half-turn which is monochromatic (green on the left helix), followed by a half-turn that has mixed colors, then one of the opposite color (red on the left helix), and then another mixed color half-turn, before repeating the pattern. This feature opens the door to a form of cohesion known as paranemic cohesion [**38**], whereby two topologically closed molecules can be made to cohere with each other by ensuring that the mixed-color domains are complementary. This principle is illustrated in Figure 3g, where a red triangle is shown to cohere to a blue triangle, even though both of them are closed. This result has been realized experimentally [**38**], and has been used intramolecularly to assemble a single-stranded octahedral structure [**30**].

5. DNA as a synthetic material

The primary advantage of using DNA in synthetic applications lies in its outstanding molecular recognition properties, enabling precise structural alignment of diverse DNA molecules that can scaffold various molecular species. DNA and its congeners appear to be unique among biopolymers in this regard. Antibodies, for example, may lead to specific binding, but the detailed geometry of antibody-antigen interactions must be worked out experimentally in every case. It is only with nucleic acids that the detailed 3D geometries are known in advance, because sticky ends form B-DNA when they bind to link two molecules together [**39**]. It is a very good approximation to assume that all sticky ends, regardless of length, have the same overall B-DNA geometry when they cohere with their complements.

The convenience of chemical synthesis [**40**] is another advantage of DNA; "vanilla" DNA of any sequence is sold by a number of vendors, and DNA synthesizers are readily capable of including a number of derivatized nucleotides, based on commercially available phosphoramidites. It is routine to synthesize DNA molecules whose lengths are 120 nucleotides or shorter. A variety of enzymes are commercially available to manipulate DNA and to trouble-shoot errors: DNA ligases catalyze the covalent joining of complexes held together by sticky-ended cohesion; exonucleases (which require an end to digest DNA) are useful in purifying cyclic target molecules from linear failure products [**41**]; restriction endonucleases are employed both to trouble-shoot syntheses and to create cohesive ends from topologically-closed species [**42**]; topoisomerases, which catalyze strand-passage reactions, can be used to correct folding errors [**43**].

There are several features of DNA that help it to perform its genetic functions. One of these is the persistence length (a measure of stiffness), about 500 Å [**28**], leading to a predictable overall structure for the short (70-100 Å) lengths typically used. In fact, it is possible to design branched species with longer persistence lengths [**44**]. Thus, the self-assembled structure of DNA constructs and some of its interactions can be taken for granted, just as one can take for granted the shape of a domino tile. Other advantages include the fact that the external code on DNA can be read, even when the double helix is intact [**45**]; thus, if DNA is used for scaffolding, absolute positions can be addressed within a pre-designed cavity. The ability to pack nanoelectronics or other scaffolded materials very tightly is likely to be aided by the high density of functional groups (every 3.4 Å or so) on DNA; consequently, DNA tile motifs with dimensions of 10-20 nm do not place an inherent limit in the close-packing of components that can be scaffolded by DNA.

Most of the work done to date utilizes DNA molecules like the molecules evolved in nature for use as genetic material. However, nanotechnology is not limited to the natural DNA molecule. A vast number of DNA analogs have been produced and analyzed for therapeutic purposes [**46**, e.g.]. This means that systems prototyped by conventional DNA may ultimately be converted to other backbones and bases, as required by specific applications. For example, it is unlikely that nanoelectronic components will be scaffolded usefully by conventional poly-anionic DNA molecules. However, there are numerous neutral analogs, most prominently PNA (a molecule with a peptide backbone to which bases are attached) [**47**], that may be much better suited to act as scaffolds for this purpose. It should be noted, however, that changing the informational polymer can lead to problems, because the helicity might be altered, with significant topological impact. For example, although DNA-DNA double helices are 10-10.5-fold helical, DNA-PNA helices are 15.5-fold helical [**48**].

6. Polyhedral catenanes

We have pointed out above that the combination of branched DNA molecules and sticky ends leads to the ability to construct N-connected objects and networks. For example, it is possible to assemble a DNA molecule whose helix axes have the connectivity of a cube or a rhombohedron [**41**]; every one of its corners consists of a three-arm branched junction. This molecule is illustrated in Figure 2d. Because the molecule has been designed to contain two turns per edge, it consists of six cyclic strands, each linked twice to its four nearest neighbors. Each of these strands corresponds to a particular face of the cube, which may be thought of as a projection down one of its fourfold axes. Had the cube been built with edges containing an odd number of double helical half-turns, it would have consisted of four strands, each being a component of six edges. Those strands would have corresponded to the hexagonal projections down the threefold symmetric body diagonals.

Each edge of the cube-like molecule contains a recognition site for a particular restriction endonuclease, so that each edge can be cleaved individually. A powerful method to demonstrate proof of synthesis for polyhedral catenanes consists of cleaving them with restriction enzymes to produce predicted catenated products that can be synthesized independently; products are analyzed by their mobility in denaturing polyacrylamide gels [**49**]. The molecule has been constructed from the linear triple catenane, corresponding to the ultimate left, front and right faces of the target object. The term "linear" is used here to indicate that the left and front faces are linked only to the front face and not to each other, in contrast, say, to the cyclic triple catenane corresponding to the front, top and right faces. One of the most robust proofs of this construction is that cleavage of the left-front and front-right edges, present in the starting material, produces the upper-back-bottom linear triple catenane. Cleavage of all the vertical edges in Figure 2d results in a single circle of appropriate size (80 nucleotides), demonstrating that a tetragonal prism has been constructed, rather than an octagonal or larger prism.

The cube-like molecule was constructed in solution, which does not afford adequate control over the synthetic process. Consequently, we developed a solid-support-based methodology, which is much more convenient [**42**]. A primary advantage is that growing molecules are isolated from one another on the support, so that the simplicity and specificity of sticky-ended ligation is not jeopardized by the presence of many copies of the same molecule in solution. In addition, the method permits control over the synthesis of individual edges of an object. As with many forms of synthesis, control derives from the use and removal of protecting groups. In this context, the protecting groups are hairpin loops. Restriction endonuclease digestion, followed by the removal of hairpin loops, generates two sticky ends that can hybridize to produce a new edge. Each cycle of the procedure creates an object that is covalently closed and topologically bonded to itself, enabling some purification to take place on the support. As with other syntheses using solid supports ([**40**, e.g.]), this methodology permits convenient removal of reagents and catalysts from the growing product. The strategy also permits the separate execution of steps involving additions and cyclizations, which are optimized under different conditions.

The efficacy of the solid support based methodology has been demonstrated by the construction of a molecule whose helix axes have the connectivity of a truncated octahedron [**50**]. This is a 14-sided polyhedron, whose ideal shape consists of six squares and eight hexagons. This geometry is illustrated in Figure 2e. The molecule was designed with two turns per edge, so it is a 14-component link, just as the cube is a six-component link; eight strands correspond to the hexagons, and six strands correspond to the squares. The square strands contain the extra arms at each vertex. Proof of synthesis is topological. First, each of the squares is shown to be in the molecule, by restricting them sequentially on their extra arms. The hexagon strands are shown to be correct by demonstrating that each of the squares is flanked by a tetracatenane of hexagons, for which a marker can be prepared independently.

It is important to reiterate that the angles flanking branched junctions are not well defined [**18, 29**]; even though the arms are quite stiff [**28**], they are flexible around the branch point, and the angles between them are variable. Thus, the geometrical designations applied to DNA polyhedra, such as a cube or a truncated octahedron, really mean that the helix axes of these objects have the same graphs as a cube or a truncated octahedron. Synthesis is both controlled and proved on the topological level, not the more rigorous geometrical level. The DX and PX motifs mentioned above are examples of geometrically well-defined DNA motifs.

7. Nucleic acid knots

There is a close relationship between knots and catenanes. Figure 4a illustrates a transformation that interconverts these two topological species. One can consider each strand of a unit tangle to have an arbitrary polarity, and then one can cleave each strand into two, so that there are two first parts and two second parts. Switching the two connections, while maintaining polarity, results in the conversion of a positive or negative node to a "zero node", where the unit tangle has been removed [**51**]. The operation is the reverse of the reciprocal exchange operation shown in Figure 3a; it is termed "Resolve" in that context. It is clear that the knot shown at the left of Figure 4a has been converted to a catenane by performing this operation on it. The right portion of the figure illustrates that performing the operation on the catenane converts it to a knot. The resolution of crossovers is found frequently in molecular biology, and is, for example, the operation performed by the enzyme T4 endonuclease VII [**52**]. In the type of chemistry shown here, the operation is usually performed on paper first, to derive a particular motif from another motif, and then the final motif is synthesized from the strands that are indicated. We will see below that one of the two end points of the robust sequence-dependent PX-JX_2 device [**53**] can be derived from the other by this resolution operation; however, their interconversion is actually achieved by the removal and replacement of particular strands, rather than by performing the strand switch operation. We have shown above in Figure 3a how to put the reverse of this operation in a DNA context [**54**]. The strand switch corresponds to the back arrows visible in panels a, c, d, e and f of that figure. Thus, the rightward arrows illustrate the node-forming operation, and the leftward arrows correspond to the strand switch operation.

Figure 4b illustrates the simplest knot, a trefoil knot, drawn in a DNA context, so that the strand has been assigned a polarity. Each of the nodes has been surrounded by a square, so that the strands of the knot form the diagonals of the square, dividing it into four regions, two between parallel strands and two between antiparallel strands. Converting the topological picture to a DNA context requires only forming a half-turn of nucleotide pairs (about six) between the antiparallel strands in each square [**55**]. The linker material between squares can be made from unstructured DNA, such as oligo-dT.

In Section 2, we introduced the idea that B-DNA produces negative nodes and Z-DNA produces positive nodes. When this idea is combined with the notion that a node or unit tangle corresponds to a half-turn of DNA, it is apparent that one can design any knot from single-stranded DNA [**55**]. We have used the branched junction motif illustrated in Figure 4b to produce a trefoil knot with negative nodes [**56**], although it is somewhat difficult to work with isolated half-turns of DNA. A much more convenient motif (Figure 4c) is one consisting of **-T-X-T-Y-T-X′-T-Y′-**, where **X** and its complement **X′** pair to form a full turn of DNA, **Y** and **Y′** do the same, and T represents a linker segment, usually composed of unstructured oligo-dT. This is the motif used to produce the first deliberate DNA trefoil knot [**57**]. The individual 104-mer strand is synthesized, and then it is cyclized with DNA ligase. We have varied the lengths of T, from a standard (used for other knots described here) of 15 nucleotides to as few as 7 nucleotides for trefoil knots [**58**].

This motif has also been employed to make figure-8 knots (sometimes known as 4_1 knots), which contain two negative nodes and two positive nodes [**59**]. This is achieved by choosing a sequence for the **X-X′** domain that is in the B conformation under ligation conditions, while the sequence of the **Y-Y′** domain is in the Z-conformation. Owing to the requirement of Z-DNA for both an appropriate sequence and Z-promoting conditions, one can produce either a trefoil knot or a figure-8 knot from the same strand, just by selecting ligation conditions. It is possible to produce sequences that form Z-DNA under stronger or weaker Z-promoting conditions [**10**]. Figure 4c illustrates that four different topologies have been produced from a strand whose **X-X′** and **Y-Y′** domains form Z-DNA in different strengths of Z-promoting conditions: A circle at low ionic strength, a trefoil knot with negative nodes at slightly higher ionic strength, a figure-8 knot at much higher ionic strength, and a trefoil knot with positive nodes when the Z-promoting reagent, $Co(NH_3)_6Cl_3$, is added [**60**].

The DNA molecule in Figure 7c adopts different conformations as a function of solution conditions; ligation to cyclize the molecule converts these conformations to topologies. Altering solutions can alter the favored conformation, and hence the favored topology. However, the actual topology is locked in by the cyclic nature of the strand; it can be changed only by passing the strand through itself. This is an operation catalyzed by type I (single-stranded) DNA topoisomerases. We have demonstrated that the interconversion of the four species in Figure 4c can be catalyzed by E. *coli* DNA topoisomerase I (topo I) and E. *coli* DNA topoisomerase III (topo III), which are both type I topoisomerases. For example, when a figure-8 knot is placed in a solution that is not conducive to the formation of Z-DNA, it is converted readily, first to the circle, and then to the trefoil knot with negative nodes, when topo I or topo III is introduced into the solution [**61**]. All the transformations are summarized in Figure 4d.

These synthetic knots are such good substrates for type I topoisomerases, that we constructed an RNA trefoil knot to seek an RNA topoisomerase activity, which was suspected, but unknown. We designed the knot similarly to the DNA trefoil knot shown in Figure 4c, and made the circle of the same sequence by altering the length of the linker used in the ligation. We discovered that E. *coli* topoisomerase III is capable of catalyzing RNA strand-passage events, although topoisomerase I does not demonstrate this activity [**62**]. Topoisomerase III was able to catalyze the interconversion of the RNA knot and the RNA circle, as well as the catenation of RNA circles. Thus, synthetic nucleic acid topology has aided in the discovery of a new activity derived from a biological system.

8. DNA junctions, antijunctions and mesojunctions

We have noted above that the relationship between a half-turn of double helical DNA and a unit tangle in a topological target suggests that it is possible to design any knot out of DNA [**55**]. If one examines the structures of various knots with this notion in mind, certain groupings of half-turn nodes become prominent [**55**]. Some are junctions, where a few half-turns all point towards a central point. These can be interpreted as branched junctions, just like the ones discussed above. For example, the arrangement labeled 33 in Figure 5a is a 3-arm branched junction, and the arrangement labeled 4_4 is a 4-arm branched junction.

However, it is possible to find other arrangements in knot diagrams. For example, the arrangement labeled 4_0 contains an empty square flanked by four half-turns. Here, the helix axes do not point to the center of the region, but are circumferential around it. This is a new type of structure, different from branched junctions, called an "antijunction" [**63**]. Figure 5a allows comparison of the branched junction, 4_4, with the antijunction, 4_0; the subscript indicates the number of helix axes that point to the center of the region flanked by the strands. It is the central local dyad axes of the half-turns that point to the center of the region in the antijunction. Other arrangements of half-turns are possible, but it is necessary to switch the orientations of the helical half-turns in pairs, so as to avoid topological ambiguities among the strands connecting them. Shown alongside the junction and the antijunction in Figure 5a are the two four-strand examples of this mixed motif, known as a "mesojunction" [**63**], ${}^{1}4_2$ and ${}^{2}4_2$; the subscripts retain their previous meaning, and pre-superscript is a label to differentiate the two species. On the left of Figure 5a the three-arm junction (3_3) is compared with the only three-strand mesojunction, 3_1.

There are no length requirements for the radial domains in these species, but the number of half-turns in the circumferential domains is critical. Owing to the difficulties of working with single half-turns of DNA, we have constructed model antijunctions and mesojunctions with an odd number of half-turns in their circumferential domains with three-half-turns [**63**]. We have also constructed model mesojunctions with two half-turns in their domains; the antijunction is unstable at this length [**64**]. These molecules are intractable. In contrast to branched junctions, which form unique species of appropriate molecular weight on nondenaturing gels, these complexes often form multimers of their target molecular weight. For example, the 3_1 mesojunction containing three half-turns in its domains forms both a monomer (3 strands, 1:2:3) and a dimer (6 strands, 1:2:3:1:2:3) under conditions where 3_3 and 4_4 are well behaved, and produce only a monomer [**63**]. The monomer and dimer are closed complexes, in which the final strand 3 is paired to the first strand 1. The antijunction is even more poorly behaved, because it forms open (noncyclic) complexes of its four strands (e.g., 1:2:3:4:1:2:3 or 1:2:3:4:1:2:3:4:1). Antijunction or mesojunction topologies arise frequently in topological design and sometimes in RNA pseudo-knot secondary structures [**64**]. Conventional branched junctions should be preferred over antijunctions or mesojunctions in topological synthetic design, because they are so much better behaved.

The failure of the antijunction to form discrete species points out an important principle in the design of nucleic acid structures, which is that it is key to take the importance of base stacking into account when designing molecular species. As noted above, any system can isomerize to produce a different arrangement from the one that has been designed if it leads to a more favorable free energy. It is also important for the molecular designer to remember that any given sample will not contain merely a single copy of the strands (probably the only ones drawn on paper), but at least 10^9 molecules, and, more likely, 10^{12} molecules, all of which are free to interact with each other.

9. Borromean rings

Borromean rings are a rich family of linked rings [**65**], named for their prominence on some of the crests of the Borromeo family, well known in the Italian renaissance [**66**]. An example is shown at the left side of Figure 5b. Although the three rings form a topologically bonded species, no pair of rings are linked together. Hence, if one of the rings were to be broken, the other two rings would fall apart. Inspection of the Borromean rings on the left of Figure 5b reveals that the signs of the three inner nodes are opposite from those of the three outer nodes. Thus, one could use the strategy described above for making knots to make the three inner nodes from half-turns of B-DNA and the three outer nodes from half-turns of Z-DNA. The three inner nodes can be regarded as flanking a three-arm B-DNA branched junction, centered in the middle of the diagram. Thus, those nodes could be coalesced to form a single molecular species.

What about the outer nodes? Whereas this is a topological diagram, it is both valid and useful to think about distorting it. One way to do so would be to wrap it around a sphere. If that were done, the three-arm B-DNA branched junction at the middle could correspond to the north pole, and the three outer nodes would each flank the south pole. Hence, they, too, could be coalesced to form a DNA branched junction. However, this junction would need to be constructed from Z-DNA, because the signs of the outer nodes are positive.

The difficulties of working with half-turns led us to substitute lengths of DNA containing three nodes for the single nodes in the synthesis. Thus, we replaced each half-turn with 1.5 turns of DNA. This operation does not change the Borromean property of the molecule, because equal numbers of positive and negative nodes have been added. So as to gain greater control, it was useful to add DNA hairpins to the equator of the molecule, between the two junctions; these hairpins do not have any effect on the topology of the molecule. These sites were used as positions where the two junctions could be ligated. Likewise, restriction sites were placed on these hairpins, so that the Borromean property could be demonstrated. The right side of Figure 5b shows that when ligated in Z-promoting conditions, the two junctions combine to form three rings with the Borromean property. The hairpins on the three rings are of slightly different lengths, although this is not indicated in Figure 5b. The different lengths permit them to be separated by denaturing gel electrophoresis. When restricted individually, only single circles are seen, and no dicatenanes are detected, demonstrating the success of the construction [**67**].

10. Topological protection

We have described the construction of two different types of topological targets: Then every nucleotide in the polyhedral catenanes is designed to base pair or to be the tight tip of a hairpin. By contrast, the knots and the Borromean rings contain long single-stranded linker regions between pairing domains. We have pointed out above that double helical DNA is quite stiff [**28**], but one has little control over the structures assumed by single-stranded DNA, which might lead to unwanted linking in the product. For example, in the first synthesis of a DNA trefoil knot, a closed species was detected with a higher mobility than the trefoil, possibly a 5_2 knot [**57**].

A solution to the problem of unwanted linking is topological protection, which entails the temporary addition of single-stranded DNA complementary to single-stranded regions. Thus, the interactions of single-stranded regions are controlled during the ligation process. Whereas these "protecting" molecules are not topologically bonded to the target molecule, they can be removed readily upon purification of the product. We have shown that this procedure is quite effective when applied to catenanes of arbitrary sequence [**34**]. Ironically, even though topological protection was devised with knots in mind, it has proved unsuccessful in that application. Indeed, attempts to apply topological protection have sometimes resulted in larger amounts of undesired products [**56**]. We ascribe this failure to the oligo-dT sequences used in the linkers; the complementary sequence, oligo-dA, is capable of forming triple helices [**68**], involving two linkers simultaneously, and leading to their linkage.

11. Two-dimensional DNA arrays from stiff tiles

The key property of the molecules generated by the motif-generating protocol discussed in Section 4 is that they are known to be relatively stiff. For example, the DX molecule has been shown to have a persistence length (a measure of stiffness) roughly twice that of conventional DNA [**44**]. Stiffness is a necessary condition for the production of periodic matter: Floppy molecules can cyclize, which will poison the crystallization. The most successful periodic [**69**, e.g.] and aperiodic [**70**] arrays have been built in two dimensions. Attempts at three-dimensional crystallization have led to crystals whose resolution is unacceptably low for structural analysis [**71**]. We should note that the criteria are different in two dimensions from what they are in three dimensions: Two-dimensional structures are usually analyzed by atomic force microscopy, whose inherent resolution is about 5-10 nanometers; three-dimensional crystals are analyzed by X-ray diffraction, whose resolution is about 1Å (0.1 nanometers); the 10 Å resolution seen in 3D would be welcomed in any two-dimensional atomic force microscopy study.

Two-dimensional arrays have been built from DX components [**69**], and components that have been extended to contain three domains, termed TX domains [**72**], cyclic 6-helix domains [**73, 74**] and planar [**75**] or nonplanar [**74, 76**] triangular motifs that have DX molecules on their edges. We have found that the double cohesion available from sticky ends in edges containing DX molecules is far more robust than the single cohesion found in simple single sticky ends [**75**]. It is possible to incorporate tiles into arrays that are marked in some fashion, either by containing an extra DNA domain, or by carrying a cargo, such as a metallic nanoparticle. This enables us to create patterns that are derived from the programmed cohesion of DNA. In addition to the periodic arrays discussed below, aperiodic arrays have also been developed that produce Sierpinski triangles [**70**], or that are capable of counting [**77**].

DX lattices have been constructed [**69, 78**] that are capable of producing patterns. As noted above, the most effective characterization of these arrays, has been achieved without ligating them, merely by visualizing them in the atomic force microscope (AFM). It is possible to include an extra domain in the DX tiles (called

DX + J), where a hairpin points up, or sometimes two different hairpins, where one points up and the other points down. These hairpins have no effect on the topology of the array, just as the hairpins added to the equator of the Borromean rings had no effect on its topology. By using the hairpin as a topographic label, it is possible to observe the structure of the array. This is particularly clear when multiple units are used; thus, instead of using the same unit repetitively, one can use a unit containing the topographic label, interspersed with one or more lacking it. When this is done, the separations between labels correspond to the repeat unit in the long direction of the DX unit; the resolution of the AFM is too low to resolve the hairpins in the direction perpendicular to the helix axes, so they just appear as stripes. When the repeating motif is changed from two units to four units (by using four different sets of sticky ends), but only one unit is labeled, the separations between the stripes doubles, confirming that the complexity of the surface features of the array can be controlled by sticky ends. This concept is illustrated schematically in Figures 5c and 5d with DAE units: Units **A** and **B**∗ (where the star represents the topographic label) are seen to form an alternating lattice whose topographic labels are separated by a double repeat. The length of the DX molecules in Figure 5c is about 16 nm. Thus, the stripes shown in the atomic force micrograph at the right are approximately 32 nm apart. When four sticky end sets are used with molecules **A, B, C** and **D**∗, the labeled tiles are separated by a quadruple repeat, or 64 nm. This is the separation seen in the AFM image shown to the right of Figure 5d. Of course, the authentic periodic repeat is two double crossover molecules in the **A-B*** array, and four double crossover molecules in the **A-B-C-D*** array.

The extent of control demonstrated over these molecules as a consequence of the sticky ends suggests that ultimately it will be possible to create intricate periodic patterns from DNA components. In a recent example, we have shown that complex components can organize a different species [**76**]. For this purpose, we use the robust 3D-DX motif [**74**]. This is a DX version of the tensegrity triangle motif first introduced by Mao and his colleagues [**79**]. The left part of Figure 5e illustrates the motif, and how the two DX domains of each direction intersect at four-arm junctions to produce its structure. The structure inherently spans 3-space, but only two of its directions of propagation are used in this experiment, leading to a two dimensional array, rather than a three-dimensional array. Note that two differently colored 3D-DX triangles are shown on the diagram, indicating molecules whose sequences are distinct, and which interact with each other in an alternating fashion. The domain that is not used for propagation is tailed with a gold nanoparticle. The central portion of Figure 5e illustrates the advantages of using two different 3D-DX units in this study. The gold nanoparticle attached to the green domain is drawn smaller than the gold nanoparticle attached to the dark blue domain. This is meant to represent the fact that the nanoparticle attached to the green domain is a 5 nm gold nanoparticle, whereas the nanoparticle attached to the dark blue domain is a 10 nm gold nanoparticle. The schematic shown in the central portion of Figure 5e indicates that the two nanoparticles ought to produce a checkerboard pattern of alternating nanoparticles in two dimensions. This prediction is borne out on the right of Figure 5e, where a checkerboard pattern of alternately large and small dots is seen in a transmission electron micrograph of the sample. Thus, in addition to generating patterns from extra DNA domains, it is possible to attach other species to DNA motifs to generate specific patterns.

12. Switchback DNA and PX DNA

The close relationship between knots and catenanes discussed in Section 7 and illustrated in Figure 4a leads one immediately to ask the question, "Is it possible to build geometrical objects from DNA knots, rather than catenanes?" Figure 6a illustrates an example of a triangle that consists of a single strand, rather than a pair of strands. However, inspection of the figure shows that the two red strands have their polarities parallel to each other, rather than antiparallel, as seen in DNA. One way to resolve this issue is to place the DNA helix axes perpendicular to the long axes of the sides of the triangle, rather than parallel with them, as might be one's first instinct. Thus, the DNA helix axes are indicated by the short double headed arrow segments perpendicular to the triangular edge, and the half-turns of base pairs are indicated by the small filled triangles in the drawing. The base pairing now occurs between antiparallel backbones. Short segments of such an arrangement produce a motif known as "Switchback DNA", also illustrated in Figure 6a.

Similar to the antijunctions discussed above, switchback DNA is not always a well-behaved motif. The reason is that, despite the elegance of the base-pairing in the design, the design of the motif fundamentally ignores the tendency for nucleic acid molecules to maximize their base stacking. Thus, when a switchback molecule is designed with an odd number of unit tangles, say five, it appears to migrate on nondenaturing gels in rough agreement with the expectation based on its molecular weight. However, when a switchback molecule is designed to contain four half-helical components, it hardly migrates at all, indicating a large molecular weight. This is because it isomerizes and forms molecules closely resembling PX molecules, which were described above in the section on motif generation. This rearrangement is also shown in Figure 6a. Indeed, the PX motif was originally discovered because switchback DNA molecules containing four half-helical units did not migrate as expected on nondenaturing gels, but seemed to form huge aggregated species. Zhiyong Shen [**80**] eventually worked out that what they were forming were infinite PX molecules. The nature of the rearrangement is shown on the right of Figure 6a, where circles represent the $5'$ ends of strands and arrowheads represent the $3'$ ends of strands. This particular molecule is self-complementary in the switchback sense, i.e., each of its four sections consists of a self-complementary hexanucleotide, so it is capable of pairing with itself, although self-complementarity is not necessary for the isomerization to the PX structure. It can be seen from the drawing how each section can pair with one other strand to produce the switchback structure or with three other strands to produce an infinite version of the PX structure. The derivation of PX DNA given in Section 4 is a generalized approach to derive the PX molecule, but it is not the way in which it was originally discovered.

13. DNA-based nanomechanical devices

There are multiple routes to obtaining nanomechanical devices from DNA-based systems, two of which will be described here. The first of these is the use of DNA structural transitions. Such transitions are well known, and they can be used to change the structure of DNA in a reliable fashion. The first device is based on the B-Z transition of DNA, discussed above, which converts right-handed B-DNA to left-handed Z-DNA [**10**]. Z-DNA has two requirements for formation: An appropriate sequence (called "proto-Z-DNA") and Z-promoting conditions; lack of either of these keeps the structure in the B-form. The need for a special sequence allows us to control the transition in space (i.e., where, and over how much Z-DNA it will occur), and the necessity of Z-promoting conditions allows us to control the transition in time (i.e., when it happens).

A device that takes advantage of the B-Z transition is illustrated in Figure 6b. It consists of two DX molecules connected by a shaft (yellow backbone) containing 20 nucleotide pairs that can undergo the B-Z transition. In the absence of Z-promoting conditions, the shaft is in the B-conformation, and both domains unconnected to the shaft are on the same side of it. However, addition of the effector, $Co(NH_3)_6^{3+}$, switches it to the Z-conformation. This in turn rotates an unconnected domain of a DX molecule to the other side of the shaft. The transition is demonstrated by a pair of dyes using fluorescent resonance energy transfer. By using 20 nucleotide pairs, the transition leads to a rotation of about 3.5 turns. It is the half-turn that leads to switching the domain on the right from the bottom to the top of the domain containing the shaft.

The problem with devices based on DNA structural transitions is that it is very hard to obtain a wide diversity of states. Using the B-Z device as an example, one could make quite a few different domains (perhaps containing a variety of other chemical species), but (except for a limited amount of chemical nuance [**60**]), one would basically be stuck with two states, the state corresponding to the device in the B conformation and a state corresponding to the device in the Z conformation. Such an approach loses one of the great strengths of DNA, its programmability by means of different sequences. This situation can be remedied by using the PX conformation and one of its topoisomers, JX_2 as the basis of the device.

Figure 6c shows the PX conformation on the left and the JX_2 topoisomer on the right. The PX molecule is drawn to show it as though two double helices, a red one and a blue one, are wrapped around each other. As noted above, the structure is a little more subtle than that, because every other half-turn along each domain consists of a red strand paired to a blue strand. The PX molecule shows that the red double helix is wrapped around the blue double helix 1.5 times. The key feature of the PX motif, of course, is that a crossover occurs between helical domains at every possible point [**80**]. The JX_2 motif differs from the PX motif in that it lacks two of the crossovers. One could imagine producing the JX_2 motif from the PX motif by the resolution operations shown in Figures 3a and 4a, but this is done only on paper, not in the laboratory. As a result of this change in topology, the red helix and the blue helix wrap around each other only once. The ends of the double helices are labeled by their colors, A and D in red and B and C in blue. Both the PX and the JX_2 structures in Figure 6c have a blue B above a red A on their left ends. However, the PX structure has a red D above a blue C on its right end, whereas the JX2 molecule has the blue C above the red D on its right end. This half-turn rotation of the two domains is the basis for the robust sequence-dependent device that has been developed [**53**].

Of course, one needs a mechanism to convert a PX motif to a JX2 motif. The mechanism is illustrated in Figure 6d, which illustrates the machine cycle. The PX state is shown on the left side of the cycle. A red strand and a blue strand have been interrupted by three half-turns of green strands. This is the region where one gains device specificity, so that a number of different devices can be addressed individually. The JX_2 state is shown on the right side of the cycle. Here, the green strands have been replaced with yellow strands. Thus, the green strands and the yellow strands constitute "set" strands that can set the state of the device.

If it were possible to replace the green strands with the yellow strands, then the two different states would be readily accessible. This replacement is achieved by a technique developed by Yurke *et al.* [**81**]. Note that both the green strands in the PX conformation contain horizontal segments; these represent a "toehold" of unpaired nucleotides. When the complete complement to the green strands is added to the solution, it pairs with the toehold, and then, through a single-stranded version of branch migration [**13**], it will displace the green set strands, leaving the naked frame shown at the top of the cycle. The complements to the set strands are shown with little black dots on their ends. These represent biotin groups, which bind very tightly to a protein called streptavidin. When magnetic beads coated with streptavidin are added to the solution, they remove the double helical set strands from the system.

Once the naked frame has been released from its set strands, it is possible to add the yellow set strands (also containing toeholds) to the solution. When this happens, the device is placed into the JX_2 state. The cycle can be completed by adding the complements to the yellow set strands (bottom right of the cycle), again liberating the naked frame. Addition of the green set strands completes the machine cycle, leading again to the PX state. Thus, it is possible to control the device by the addition of a series of setting strands that change its topology, because the green set strands produce crossovers, and the yellow set strands produce only simple duplex motifs.

The operation of the device has been shown on gels, but gels are not a very satisfying way to demonstrate the states of a nanomechanical device. Figure 6e illustrates a more convincing experiment, which involves atomic force microscopy. The PX-JX_2 device has been used to connect a number of DNA half-hexagon trapezoids. When the devices are all in the PX state, they produce trapezoids that all point in the same direction. However, when switched to the JX_2 state, the trapezoids alternately point in opposite directions. The parallel nature of the trapezoids is shown in the atomic force micrographs in the square near the PX state, and the alternating orientation of the trapezoids is shown in the square near the JX_2 state. Thus, it is possible to visualize the operation of the PX-JX_2 nanomechanical device.

In the years since the PX-JX_2 device was first developed, its use has been extended markedly. Two PX-JX_2 devices have been coupled to other DNA motifs, so as to build a machine that translates the sequences in set strands into polymer assembly instructions [**82**]. In a second advance, RNA has been used to control the DNA-based PX-JX_2 device [**83**]. The PX-JX_2 device has been incorporated into a cassette, so that it can be inserted into a 2D DNA array [**84**]. Recently, a third state of the device has been produced from the frame by forming a contracted state, so that this can now be used as the basis for a nanomechanical device with three robust states.

14. Future topological targets

What does the future hold for synthetic single-stranded DNA topology? It is always risky to suggest the next things to occur in any field. I will venture to guess only what will happen in two areas. The first is in the domain of automatic weaving and braiding on the nanoscale. Woven molecular structures are extremely hard to produce in a deliberate fashion now. Interwoven structures made exclusively from right-handed DNA, such as the one shown in Figure 2c, are certainly possible, but most woven structures require an alternation of negative and positive nodes. As noted above, one can get positive nodes from DNA by using Z-DNA, but the range of sequences is extremely limited, and the conditions for forming Z-DNA are special, and not necessarily compatible with other features of the solution conditions that may be desired. On the other hand, molecular systems that could automatically weave themselves would be extremely powerful, and would be likely to lead to very strong materials. The solution to this problem may now be at hand, because of several investigators [**85**, e.g.] have built nucleoside phosphoramidites from deoxyribose sugars of the opposite chirality from that found in naturally-occurring DNA. Consequently, left-handed segments of arbitrary sequence will soon be available for building self-weaving materials. It is probably worth pointing out that cylindrically woven systems are actually easier to construct than closed systems that are planar (resembling the woven patterns in Roman mosaics), because of issues involving the polarity of the DNA backbone. Nevertheless, this is a minor technical annoyance, and certainly not a fundamental stumbling block.

As interesting and intricate as DNA is, there are many purposes for which other polymers would be more desirable. Ideally, one would like to be able to control the topology of any polymer, not just DNA. The chain mail shown in Figure 2b is a very tempting target, but is certainly not even remotely accessible in any polymer other than DNA or one of its congeners. One solution to this issue is to construct ladder polymers of DNA and another polymer, wherein the DNA directs the topology of the other polymer. Removal of the DNA would then lead to a topological target in the other polymer. Steps in this direction have been taken by James Canary's laboratory [**86**]. Nucleoside phosphoramidites have been developed that contain nylon components (dicarboxylic acids and diamino groups) that are pendent from the $2'$ positions of alternating nucleosides. The carboxyl groups and the amino groups can be coupled successfully when the DNA is paired with an appropriate complement. Consequently, the topology of the DNA can be transferred to the nylon backbone. To date, no closed molecule or molecular complex, such as a catenane or a knot, has been built in this system, but it is unlikely to be a very long time before that goal is achieved.

15. Synthetic single-stranded DNA topology and structural DNA nanotechnology

The plectonemic (interwrapped) nature of the DNA double helix makes it a tractable molecule for experiments in molecular topology. This is a very rich vein for the exploration of the topological properties of matter. In this chapter, we have tried to illuminate some of the techniques by which the single-stranded topology of DNA can be directed in synthetic molecules. Catenanes, knots and Borromean rings are available from simple protocols, and it is to be hoped that the physical properties associated with complex topologies [**87**] will become available through the medium of DNA construction. In addition to the generation of targets as ends in themselves, we have noted that a topological approach has led to the characterization of the sign of the parallel Holliday junction crossover [**34, 35**], and to the discovery of RNA topoisomerase activity [**62**].

Are there any other possible uses for the construction of complex topological species? One possible application is in the mass production of DNA polyhedral catenanes by biological means, such as the polymerase chain reaction (PCR) [**88**] or by production in *vivo.* Shih and his colleagues [**30**] have indeed utilized DNA topology to produce a single-stranded DNA octahedron, although he used five small helper strands to simplify the process. He built the octahedron from edges primarily consisting of PX DNA. Using the PX cohesion shown in Figure 3g, he demonstrated that he could reproduce the strands that he needed biologically, and that they could fold into the target octahedral structure.

Another major result involving single-stranded DNA is DNA origami, developed recently by Paul Rothemund [**89**], based on the long-single-stranded precedents of Shih *et al.* [**30**] and Yan *et al.* [**90**]. DNA origami is the name given to the use of a long single-stranded viral strand to make convenient shapes. For example, Rothemund has worked out methods to combine about 250 smaller "staple" strands with M13 viral DNA to make shapes ranging from rectangles and triangles to smiley faces. By adding extra domains to the staple strands, he has also been able to generate patterns that include the a map of the western hemisphere.

I have pointed out that topological constructions are intimately related to geometrical stick figures built from DNA [**91**]. Those molecules are designed with the aim of controlling the structure of matter on the nanometer scale; hence their geometries are ultimately of key interest. The goals of such catenane constructions include the rational design of periodic matter [**14, 92**], nanoscale mechanical devices [**93**] and biochips for computation on the smallest possible scales [**94**]. Some of the molecules to be associated with DNA will be pendent from it, hanging off every residue; if that is the case, then the topology of the pendent molecules will reflect that of the DNA scaffold. In the construction of two-dimensional DNA arrays, certainly the most successful prototyping of designed periodic matter from DNA, the topological designs have been key to the assembly of the molecules. The importance of topology is also key in the current attempts to extend periodic matter to three dimensions.

When working with a well-structured polymer, like DNA or RNA [**95**], its topological properties are the first ones that can be characterized: How is it linked? How is it branched? These properties are present, even if the polymer is not completely rigid. The area of structural DNA nanotechnology is developing from a topologically-based discipline to a geometrically-based discipline, as rigid motifs, such as DX or PX molecules are introduced. It is to be expected that this increase in geometrical definition will heighten the importance of topological considerations in nanoconstruction: We expect that rigid components will enable us to achieve greater control over topological targets, and greater control over their analysis. As we achieve control in both these areas, synthetic DNA topology will continue to mature.

Figure Legends

Figure 1. DNA nodes, branched junctions and cohesion. (a) The Relationship Between Crossings and DNA Double Helices. The top portion of this drawing illustrates that two polar strands can cross each other in two different ways that are mirror images of each other. The negative crossing, or node, is shown on the left and the positive node is shown on the right. The bottom portion shows that the crossings generated by B-DNA produce negative nodes, whereas the crossings generated by Z-DNA produce positive nodes. Z-DNA contains a zig-zag backbone, so its structure is somewhat different from that of B-DNA. Nevertheless, the crossings are positive. (b) A Stable DNA Branched Junction. The junction shown is composed of four strands of DNA, labeled with Arabic numerals; arms are labeled with Roman numerals. The $3'$ end of each strand is indicated by the half-arrows. Each strand is paired with two other strands to form a double helical arm. When the four different strands are added to the same solution, they bind to each other to form the double helical segments shown. (c) Highly Branched DNA Junctions. The top row contains a 5-arm junction and a 6-arm junction. The bottom row contains an 8-arm junction and a 12-arm junction. The same conventions apply as in (b). The number of arms corresponds to the ultimate maximum connectivity of any object or network in which the junctions are components. (d) Two linear double helical molecules of DNA are shown at the top of the drawing. The antiparallel backbones are represented by the thick lines that terminate in half-arrows that indicate the $5' \rightarrow 3'$ directions of the backbones. The right end of the left molecule and the left end of the right molecule have single-stranded extensions ("sticky ends") that are complementary to each other. The middle portion shows that, under the proper conditions, these bind to each other specifically by hydrogen bonding. The bottom of the drawing shows that they can be ligated to covalency by the proper enzymes and cofactors.

Figure 2. The combination of branched junctions and sticky ends. (a) Formation of a two-dimensional lattice from an immobile junction with sticky ends. X is a sticky end and X' is its complement. The same relationship exists between Y and Y'. Four of the monomeric junctions on the left are complexed in parallel orientation to yield the structure on the right. X and Y are different from each other, as indicated by the pairing in the complex. As shown in Figure 1d, ligation by DNA ligase can close the gaps left in the complex. The complex has maintained open valences, so that it could be extended by the addition of more monomers. (b) and (c) Topological Consequences of Ligating DNA Molecules Containing Even and Odd Numbers of DNA Half-Turns in Each Edge. These diagrams represent the same ligation shown in (a), although a little larger. Here, however, they indicate the plectonemic winding of the DNA, and its consequences. The DNA is drawn as a series of right-angled turns. In (b) each edge of the repeating square contains two turns of double helix. Strands are drawn in red and blue, leading to cyclic strands of alternate colors. In (c), each edge of the repeating square contains 1.5 turns of DNA. Therefore, the strands do not form cycles, but extend infinitely in a warp and weft interwoven meshwork, with red lines on the lower-right, upper left diagonal, and blue lines on the lower-left, upper right diagonal. (d) A Hexacatenane DNA molecule whose helix axes have the connectivity of a cube. The molecule shown consists of six cyclic strands in six different colors, with the bases shown in white.

Each edge contains 20 nucleotide pairs of DNA, so we expect that their lengths will be about 68 Å. Each face corresponds to a cyclic single strand. For example, the front face corresponds to the red strand, and it is linked twice to the green, cyan, magenta and dark blue strands that flank it, and only indirectly linked to the yellow strand at the rear. (e) A 14-catenane DNA molecule whose helix axes have the connectivity of a truncated octahedron. Each edge here is also two turns long, and the same conventions apply as in (d). There are eight hexagonal faces and six square faces; the squares are all colored red. Note that although the truncated octahedron is a 3-connected object, it has been made from 4-arm junctions, so there is a hairpin at each vertex, all emanating from the red square strands.

Figure 3. Motif generation by reciprocal exchange. (a) Reciprocal exchange. A red strand and a blue strand are shown a left. Following reciprocal exchange, they have been converted to a red-blue strand and a blue-red strand forming a node. The reverse operation, creating the red strand and blue strand and destroying the node is resolution. (b) A kolam from the floor of the Meenakshi temple in Madurai. Reciprocal exchanges are evident in both directions in this pattern. (c) Reciprocal exchange between two double helices. Exchange is shown between strands of the same polarity in two double helices; $3'$ ends are indicated by arrows. A four-stranded structure results from this operation, joining the two double helices. The strands that have undergone reciprocal exchange are colored green. The twofold symmetry of the structure is indicated by the double-headed arrow. (d) Double reciprocal exchange between two double helices. The two exchanges again occur between strands of the same polarity, colored red on both sides of the reaction. The other strands are colored green. (e) Formation of PX DNA by reciprocal exchange everywhere between strands of the same polarity. The result of this operation is two double helices wrapped around each other. Each helical domain consists of a green half-turn, a green-red half-turn, a red half-turn and then a red-green half turn, before repeating. (f) Double reciprocal exchange between strands of opposite polarity. A different structure results from (d), containing five strands, rather than four. Two unperturbed helical strands are colored green, two red strands cross over between domains and then reverse direction, and then a fifth cyclic blue strand results in the center of the motif. (g) PX cohesion. Two topologically closed triangles, one blue and one red, cohere by forming a PX motif. The cohesion that results is paranemic.

Figure 4. DNA knots. (a) Strand switch to remove nodes in knots and catenanes. On the left is a 5-noded knot (51), with its polarity indicated by arrowheads. Passing to the middle, one strand switch has been performed, converting the knot to a catenane, drawn with lines of two different colors, red and blue. On the right, another strand switch has been performed, making a new 3-noded knot. (b) The relationship between a half-turn and a node. A trefoil knot has been drawn in red; its polarity is shown by the arrowheads along the knot. A square box has been drawn in dark green about each node, so that the strands of the knot divide the boxes into four regions, two between antiparallel strands, two between parallel strands. A half-turn of base pairs is drawn between antiparallel strands; the helix axes are shown as double-headed light green arrows and dyad axes normal to them are represented by magenta lines ending in two ellipses. (c) The synthetic scheme used to produce four topologies. The top of this synthetic scheme indicates the molecule from which the target products are produced. The four pairing regions, X and its complement **X**′, **Y** and its complement **Y**′ are indicated by the bulges from the square. The 3′ end of the molecule is denoted by the arrowhead. The four independent solution conditions used to generate the target products are shown below the basic structure. The molecular topology of the products is shown at the bottom of the panel; signs of nodes are indicated. (d) Transformations between knots catalyzed by topoisomerase I. The trefoil knot with positive nodes of (c) is shown at the left. As the Z-promoting conditions are decreased, the products are the figure-8 knot and then the trefoil knot with negative nodes, proceeding from left to right. The circular intermediates between each of these knots are also indicated.

Figure 5. Topological and periodic structures. (a) Schematic drawings of 3-strand and 4-strand junctions, antijunctions and mesojunctions shown as the helical arrangements that can flank a triangle or a square. Each polygon is formed from differently colored strands of DNA that extend beyond the vertices in each direction. The arrowheads indicate the 3′ ends of the strands. The vertices correspond to the nodes formed by a half-turn of double helical DNA. Base pairs are represented by aqua lines between antiparallel strands. Thin double-headed gold arrows perpendicular to the base pairs represent the axis of each helical half-turn. The rust-colored lines perpendicular to the helix axes terminating in ellipses represent the central dyad axes of the helical half-turns. The complexes 3_3 and 4_4 correspond to conventional branched junctions. The complex 4_0 is a 4-strand antijunction. The complexes 3_1, ${}^{1}4_2$ and ${}^{2}4_2$ are mesojunctions, which contain a mix of the two orientations of helix axes. (b) Borromean rings. Traditional Borromean rings are drawn on the left in red, green and blue. Note that no two rings are linked. The synthesis is shown on the right: A B-DNA branched junction (left) and a Z-DNA junction (right) are shown, each with 1.5 turns per arm; the signs of nodes are noted. The three strands are drawn with lines of different colors. At the ends of each strand are partial hairpins (one almost complete, one that supplies the missing portion to another incomplete hairpin). When ligated in Z-promoting conditions, the Borromean arrangement on the right is obtained. (c) and (d) Two-dimensional arrays constructed from DX and DX+J motifs. Two arrays are shown, one (c) containing two components, A and B*, and a second (d) containing four components, A, B, C and D*. Each of the starred components contains one or two hairpins perpendicular to the plane of the array. The double crossover molecules are represented as two closed figures connected by two short lines. The sticky ends are drawn schematically as complementary geometrical shapes, representing Watson-Crick complementarity. The horizontal repeat is two units in the top array and four in the bottom array; the vertical repeat is a single unit in both arrays. The perpendicular hairpins are visible as stripes when this array is examined in the atomic force microscope, as seen to the right of each array. (e) Organizing gold nanoparticles using a robust DNA motif. Four copies of the motif used are shown on the left, two each drawn in different colors, to indicate that two separate molecules form the two-dimensional lattice. A simpler representation of each molecule is shown in the center, one with a small (5 nm) gold nanoparticle hung from one domain and one with a large (10 nm) gold nanoparticle hung from the analogous domain; a schematic of the checkerboard two-dimensional array is shown to the right. A transmission electron micrograph of the actual array is shown on the far right of the panel.

Figure 6. PX DNA and nanomechanical devices. (a) The left part of this panel shows the origin of switchback DNA in a knot that forms a triangle. The single strand of DNA is shown in red, and its polarity is indicated by arrowheads along it. It is clear that its two cycles are parallel to each other. The antiparallel base pairing in a half-turn in each edge is shown as filled blue sections, and helix axes perpendicular to the pairing are indicated by arrows pointing in opposite directions. The right part of the panel shows a switchback molecule (SB) with four half-turn domains, each numbered and colored differently. The $5'$ end of each strand is indicated by a filled circle and the $3'$ end by an arrowhead. Below the switchback molecule is the structure of the PX molecule to which it isomerizes to maximize its base stacking. The helix axes are indicated in both molecules by arrows pointing in opposite directions. The nicks in the PX molecule have been drawn to be at the edges for clarity, but they are probably closer to the crossovers. (b) A DNA nanomechanical device based on the B-Z transition. The device consists of two DX molecules (drawn red and blue) connected by a shaft of proto-Z DNA (yellow). The green and magenta circles indicate a pair of dyes for fluorescence resonance energy transfer measurements that indicate that the device changes state. When $Co(NH_3)^{3+}$ is added to the solution, the proto-Z DNA converts from B-DNA to Z-DNA and the domain on the right changes its position from the bottom to the top of the shaft. (c) PX and JX_2 motifs. The motif on the left is PX DNA and on the right is a topoisomer of PX, JX_2, which lacks two of the crossovers of the PX molecule. As a consequence, the right side of the JX_2 molecule is rotated by a half-turn from the right side of the PX molecule. (d) The machine cycle of the PX-JX_2 device. Starting from the PX state on the left, the green set strands are removed by the addition of their complete complements, including their toeholds; the duplexes are removed from the solution by the binding of magnetic streptavidin beads to biotins (shown as black dots) on the complement strands. Yellow set strands can then be added to the naked frame (top) to produce the JX_2 state (right). The cycle is completed by removing the yellow set strands and replacing them with the green set strands. (e) Atomic force microscopic demonstration of the device. The schematic at left shows that the device is used to connect a series of DNA trapezoids. In the PX state the trapezoids are parallel to each other, but they form a zigzag pattern in the JX_2 state. The atomic force micrographs at the right demonstrate these patterns.

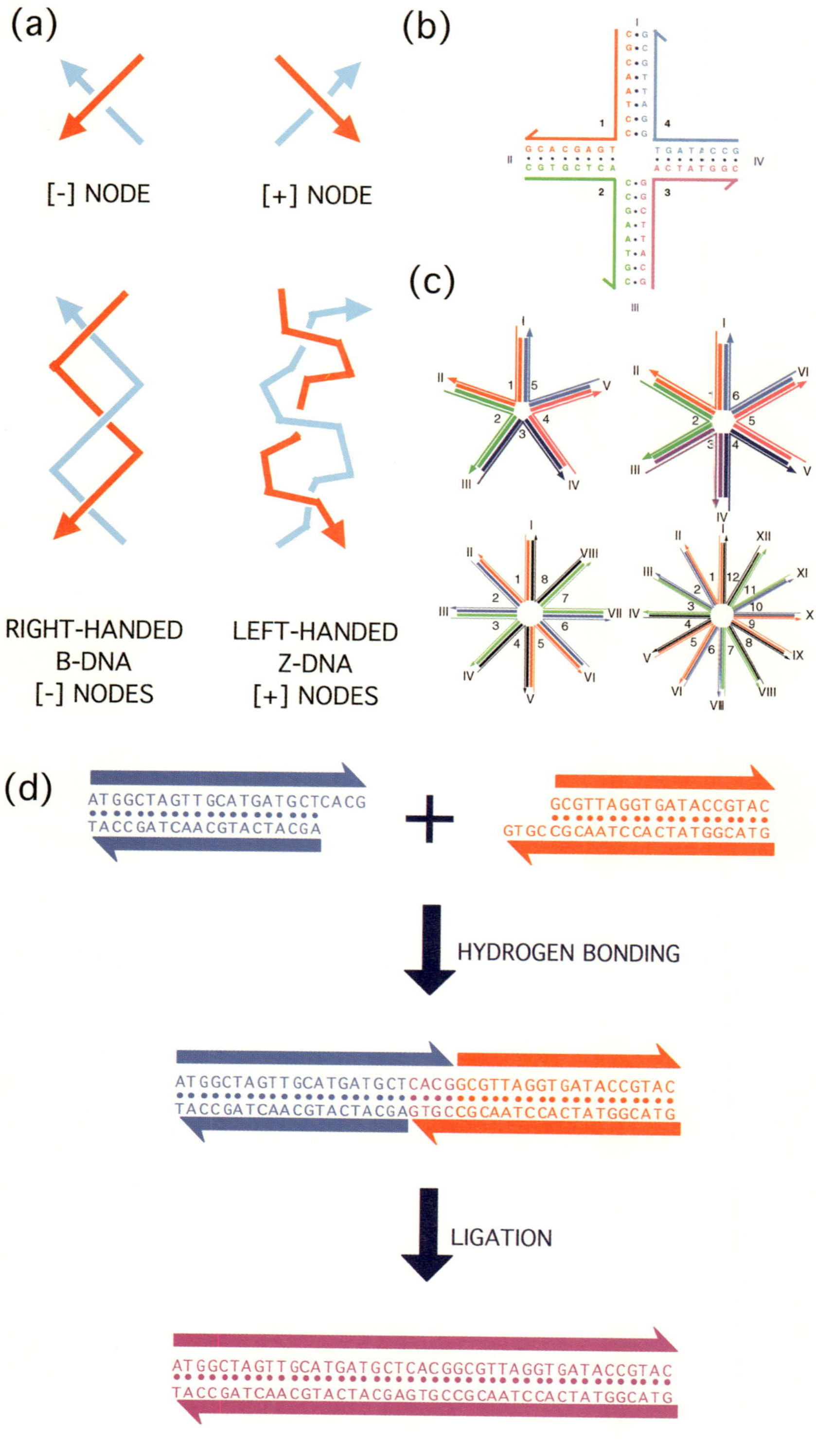
(a)
[-] NODE
[+] NODE
RIGHT-HANDED
B-DNA
[-] NODES
LEFT-HANDED
Z-DNA
[+] NODES
(b)
(c)
(d)
ATGGCTAGTTGCATGATGCTCACG
TACCGATCAACGTACTACGA
GCGTTAGGTGATACCGTAC
GTGCCGCAATCCACTATGGCATG
HYDROGEN BONDING
ATGGCTAGTTGCATGATGCTCACGGCGTTAGGTGATACCGTAC
TACCGATCAACGTACTACGAGTGCCGCAATCCACTATGGCATG
LIGATION
ATGGCTAGTTGCATGATGCTCACGGCGTTAGGTGATACCGTAC
TACCGATCAACGTACTACGAGTGCCGCAATCCACTATGGCATG

FIGURE 1

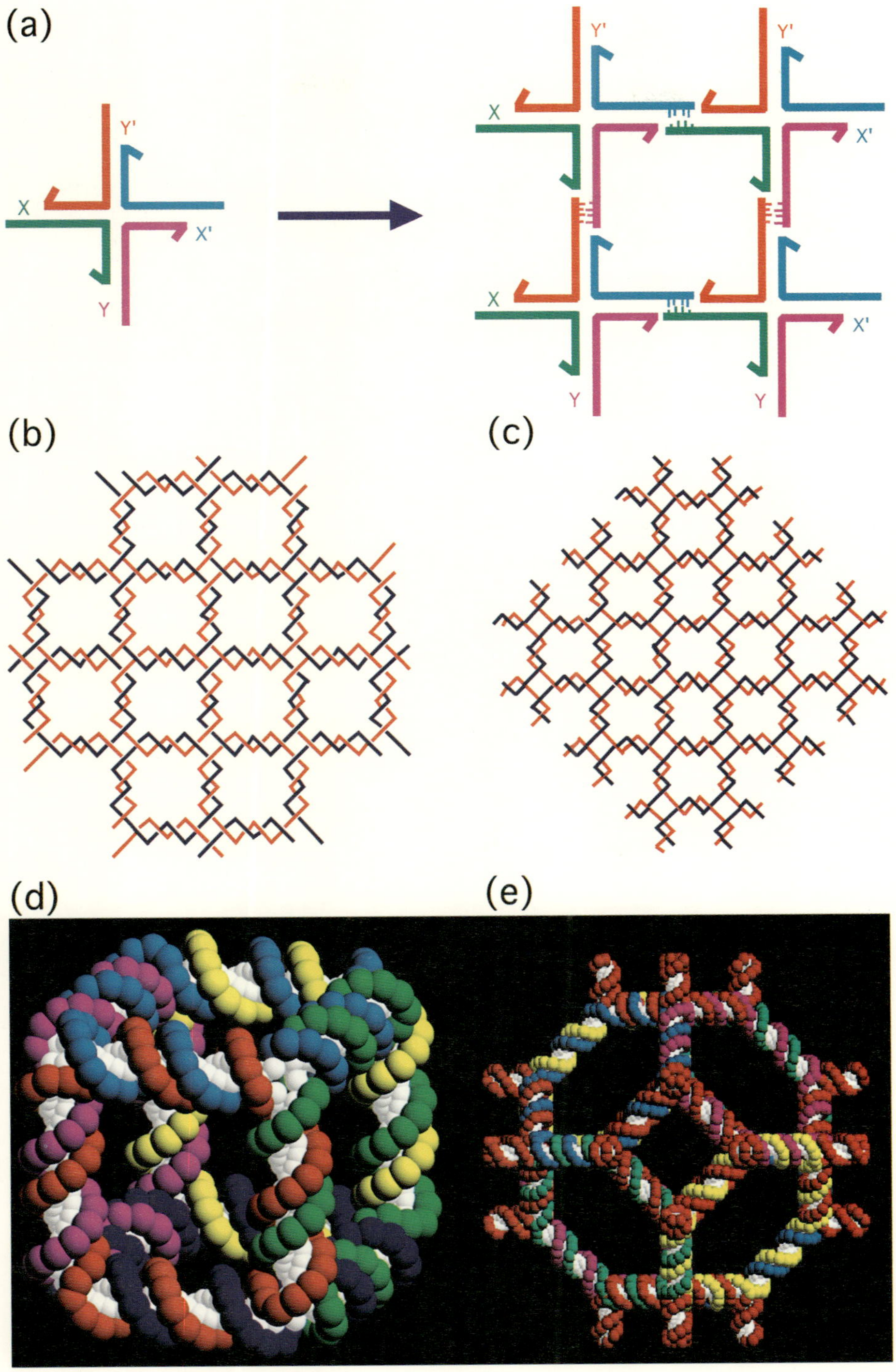

FIGURE 2

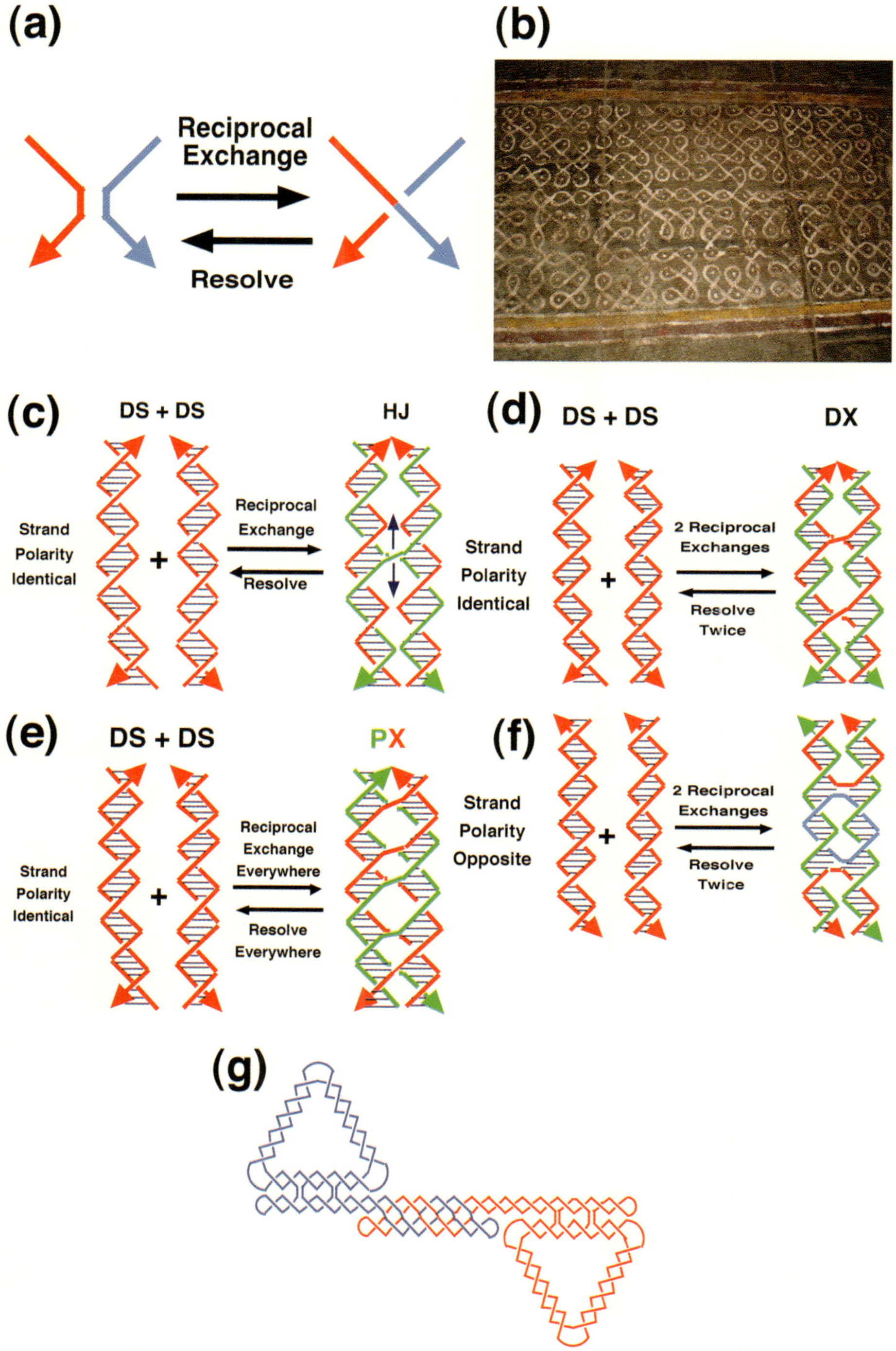
(a)
Reciprocal Exchange
Resolve
(b)
(c)
DS + DS
HJ
Strand Polarity Identical
Reciprocal Exchange
Resolve
(d)
DS + DS
DX
Strand Polarity Identical
2 Reciprocal Exchanges
Resolve Twice
(e)
DS + DS
PX
Strand Polarity Identical
Reciprocal Exchange Everywhere
Resolve Everywhere
(f)
Strand Polarity Opposite
2 Reciprocal Exchanges
Resolve Twice
(g)

FIGURE 3

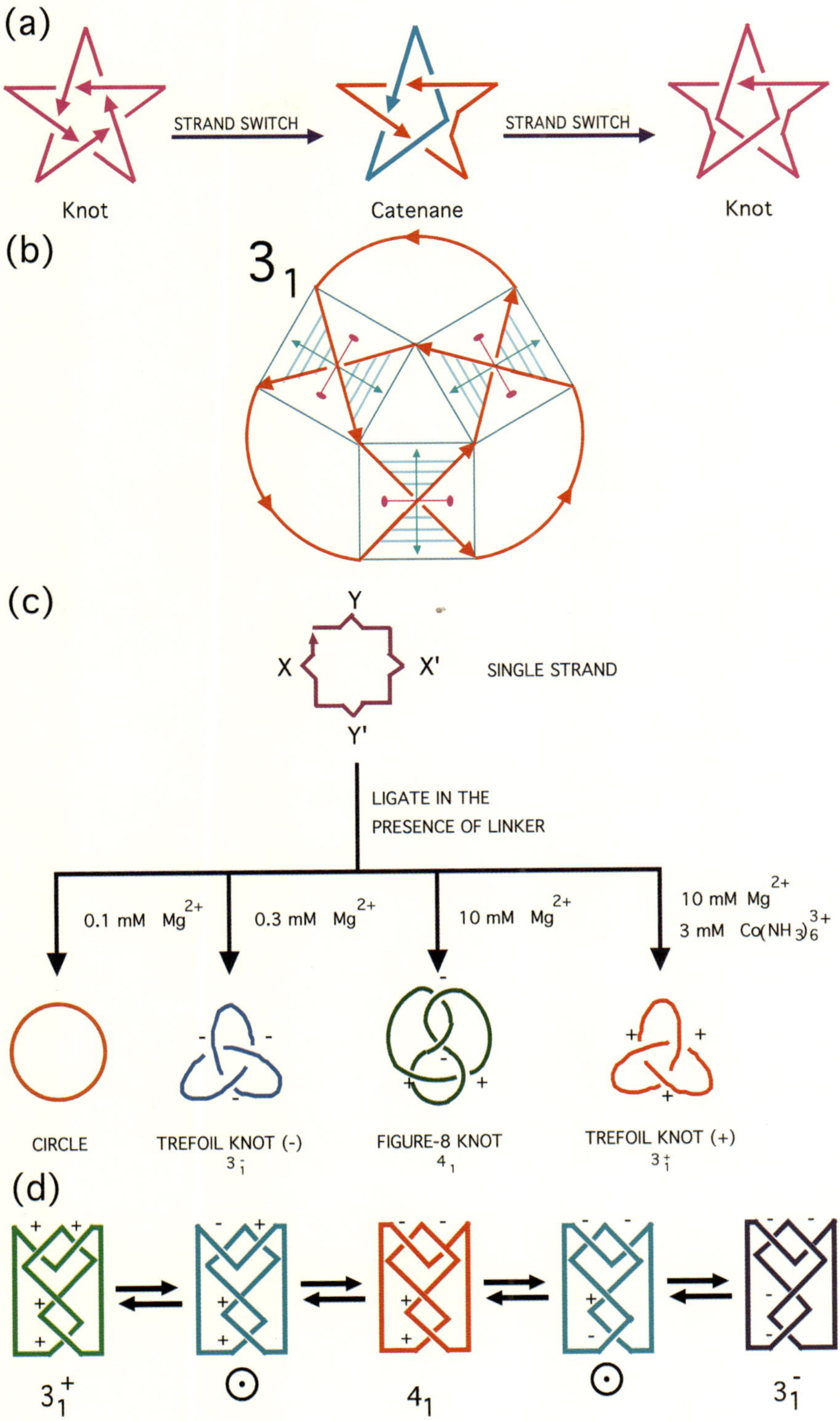
(a)
STRAND SWITCH
STRAND SWITCH
Knot
Catenane
Knot
(b)
3 1
(c)
Y
X
X'
Y'
SINGLE STRAND
LIGATE IN THE
PRESENCE OF LINKER
0.1 mM Mg 2+
0.3 mM Mg 2+
10 mM Mg 2+
10 mM Mg 2+
3 mM Co(NH 3)6 3+
CIRCLE
TREFOIL KNOT (-)
FIGURE-8 KNOT
TREFOIL KNOT (+)
(d)

FIGURE 4

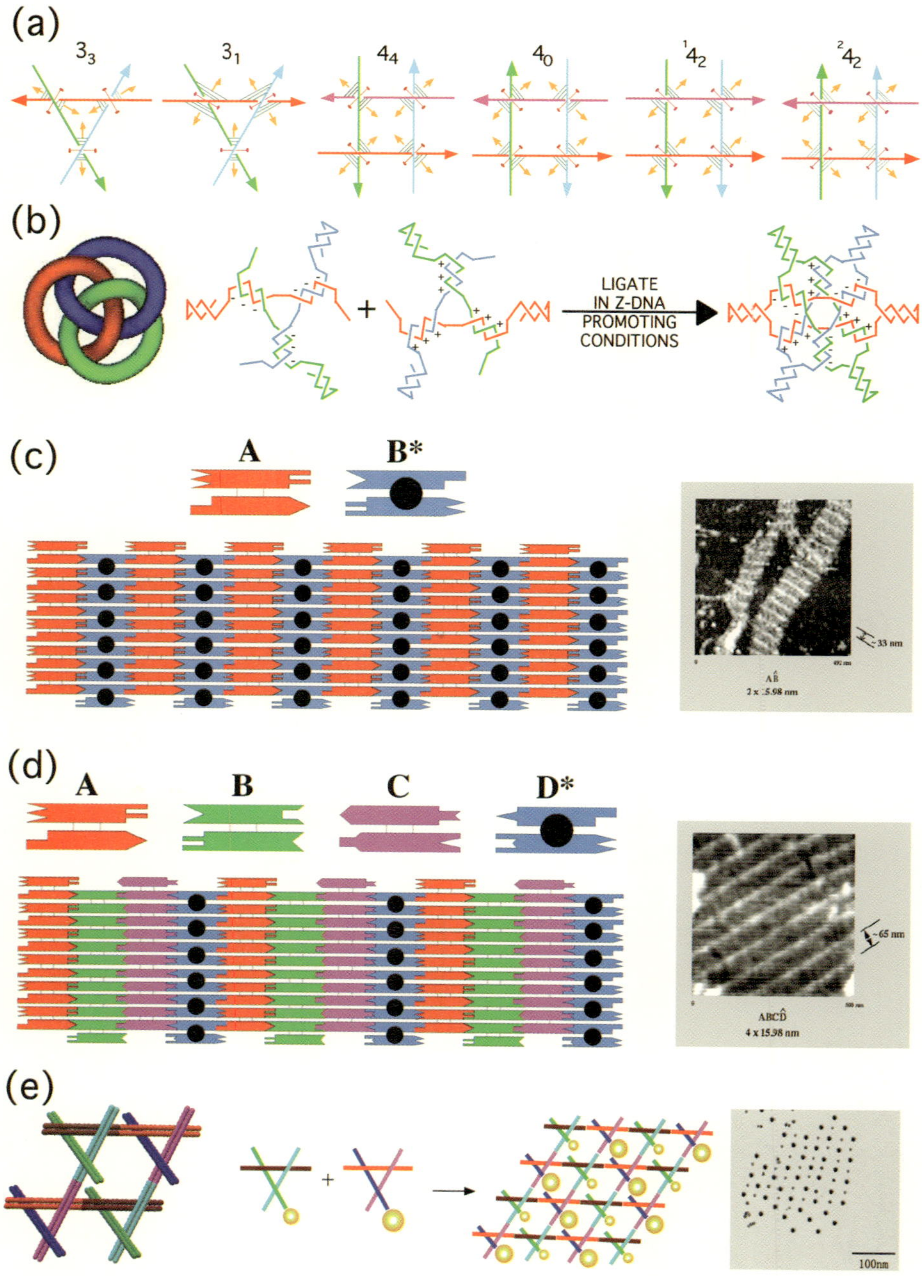
(a)
3₃
3₁
4₄
4₀
¹4₂
²4₂
(b)
LIGATE
IN Z-DNA
PROMOTING
CONDITIONS
(c)
A
B*
~33 nm
(d)
A
B
C
D*
~65 nm
4 x 15.98 nm
(e)
100nm

FIGURE 5

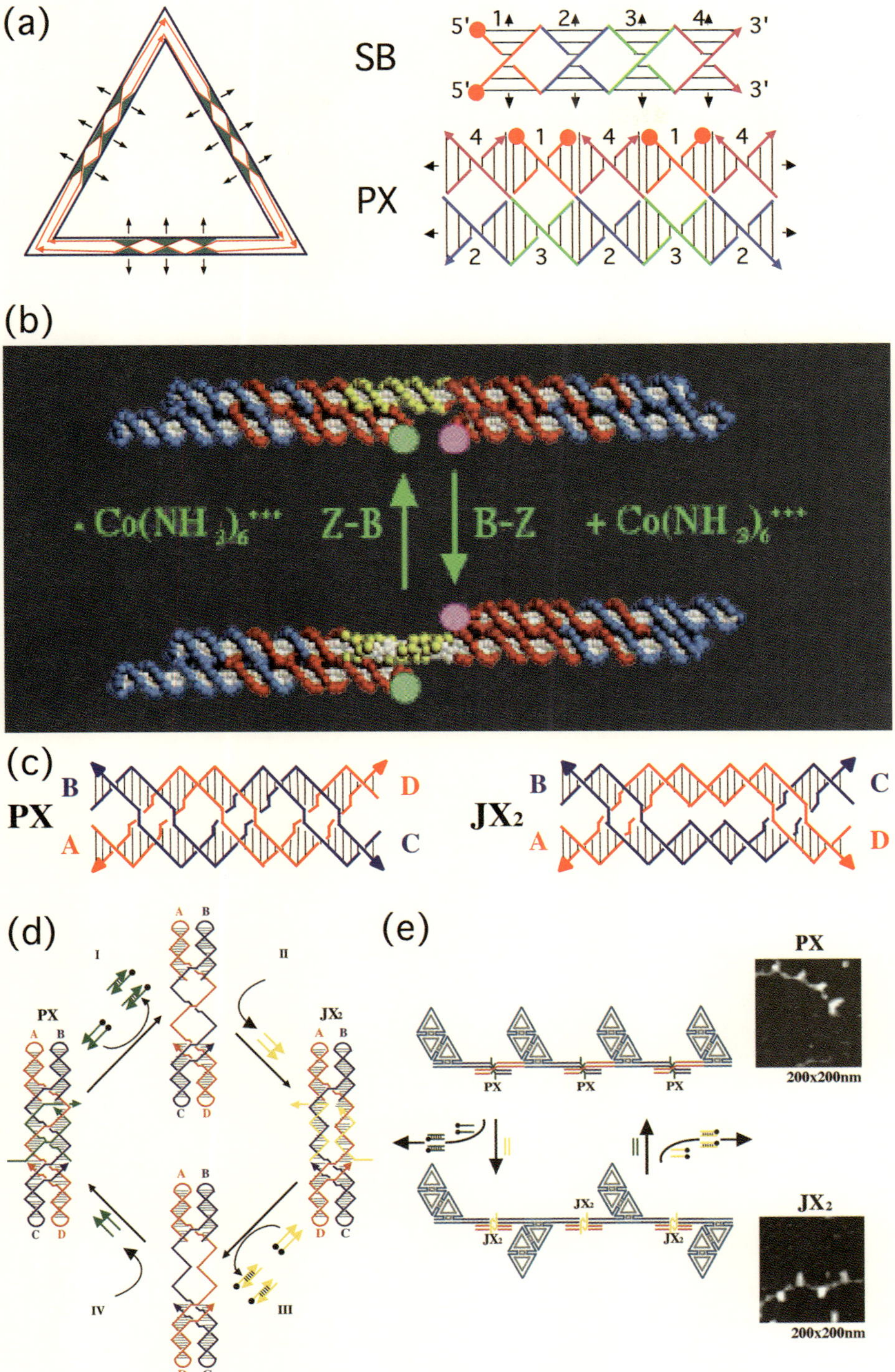
(a)
SB
5'
3'
1
2
3
4
PX
(b)
- Co(NH3)6+++
Z-B
B-Z
+ Co(NH3)6+++
(c)
PX
A
B
C
D
JX2
(d)
I
II
III
IV
(e)
200x200nm

FIGURE 6

References

[1] J.D. Watson and F.H.C. Crick, *A structure for deoxyribose nucleic acid*, Nature **171** (1953), 737-738.

[2] D. W. Sumners, *Untangling DNA*, Math. Intelligencer **12** (1990), 71-80.

[3] J. P. Sauvage, *Interlacing molecular threads on transition metals*, Accts. Chem. Res. **23** (1990), 319-327.

[4] D.B. Amabilino and J.F. Stoddart, *Interlocked and intertwined structures and superstructures*, Chem. Rev. **95** (1996), 2725-2828.

[5] K.S. Chichak, S.J. Cantrill, A.R. Pease, S.H. Chiu, G.W.V. Cave, J.L. Atwood, and J.F. Stoddart, *Molecular Borromean rings*, Science **304** (2004), 1308-1312.

[6] R. Jäger and F. Vögtle, *A new synthetic strategy towards molecules with mechanical bonds*, Angew. Chem. Int. Ed. Engl. **36** (1997), 930-944.

[7] N.R. Cozzarelli and J.C. Wang, *DNA Topology and its Biological Effects.* Cold Spring Harbor Laboratory Press, 1990.

[8] A.D. Bates and A. Maxwell, *DNA Topology.* IRL Press at Oxford University Press, Oxford, 1993.

[9] R. Wing, H. Drew, T. Takano, C. Broka, S. Tanaka, K. Itakura, and R.E. Dickerson, *Crystal structure analysis of a complete turn of B-DNA*, Nature **287** (1980), 755-758.

[10] A. Rich, A. Nordheim, and A.H.-J. Wang, *The chemistry and biology of left-handed Z-DNA*, Ann. Rev. Biochem. **53** (1984), 791-846.

[11] C. Mao, W. Sun, Z. Shen, and N.C. Seeman, *A DNA nanomechanical device based on the B-Z transition*, Nature **397** (1999), 144-146.

[12] R. Holliday, *A mechanism for gene conversion in fungi*, Genet. Res. **5** (1964), 282-304.

[13] P. Hsieh and I.G. Panyutin, *DNA branch migration*, In Nucleic Acids & Mol. Biol., eds., F. Eckstein, D.M.J. Lilley, **9**, 42-65, Berlin: Springer-Verlag, 1995.

[14] N.C. Seeman, *Nucleic acid junctions and lattices*, J. Theor. Biol. **99** (1982), 237-247.

[15] N.C. Seeman, *De novo design of sequences for nucleic acid structural engineering.* J. Biomol. Struct. & Dyns. **8** (1990), 573-581.

[16] N.C. Seeman and N.R. Kallenbach, *DNA branched junctions*, Ann. Rev. Biophys. Biomol. Struc. **23** (1994), 53-86.

[17] D.M.J. Lilley and R.M. Clegg, *The structure of the four-way junction in DNA*, Ann. Rev. Biophys. Biomol. Str. **22** (1993), 299-328.

[18] R.-I. Ma, N.R. Kallenbach, R.D. Sheardy, M.L. Petrillo, and N.C. Seeman, *Three arm nucleic acid junctions are flexible*, Nucl. Acids Res. **14** (1986), 9745-9753.

[19] N.R. Kallenbach, R.-I. Ma, and N.C. Seeman, *An immobile nucleic acid junction constructed from oligonucleotides*, Nature **305** (1983), 829-831.

[20] Y. Wang, J.E. Mueller, B. Kemper, and N.C. Seeman, *The assembly and characterization of 5-Arm and 6-Arm DNA junctions*, Biochem. **30** (1991), 5667-5674.

[21] X. Wang and N.C. Seeman, *The assembly and characterization of 8-arm and 12-arm DNA branched junctions*, J. Am. Chem. Soc. **129** (2007), 8169-8176.

[22] N.C. Seeman and N.R. Kallenbach, *Nucleic acid junctions: A successful experiment in macromolecular design.* In: Molecular Structure: Chemical Reactivity and Biological Activity, eds., J.J. Stezowski, J.L. Huang, M.C. Shao, Oxford University Press, Oxford, pp. 189-194, 1988.

[23] S.N. Cohen, A.C.Y. Chang, H.W. Boyer, and R.B. Helling, *Construction of biologically functional bacterial plasmids in vitro*, Proc. Nat. Acad. Sci. (USA) **70** (1973), 3240-3244.

[24] A.F. Wells, *Three-Dimensional Nets and Polyhedra,* John Wiley & Sons, New York, 1977.

[25] R. Williams, *The Geometrical Foundation of Natural Structure*, Dover, New York, 1979.

[26] R.M. Dirks, M. Lin, E. Winfree, and N.A. Pierce, *Paradigms for nucleic acid design*, Nucl. Acids Res. **32** (2004), 1392-1403.

[27] D. Voet and A. Rich, *The crystal structures of purines, pyrimidines and their intermolecular complexes*, Prog. Nucl. Acid. Res. Mol. Biol. **10** (1970), 183-265.

[28] P.J. Hagerman, *Flexibility of DNA*, Ann. Rev. Biophys. & Biophys. Chem. **17** (1988), 265-286.

[29] M.L. Petrillo, C.J. Newton, R.P. Cunningham, R.-I. Ma, N.R. Kallenbach, and N.C. Seeman, *Ligation and flexibility of four-arm DNA junctions*, Biopolymers **27** (1988), 1337-1352.

[30] W.M. Shih, J.D. Quispe, and G.F. Joyce, *DNA that folds into a nanoscale octahedron*, Nature **427** (2004), 618-621.

[31] R.P. Goodman, I.A.T. Schaap, C.F. Tardin, C.M. Erben, R.M. Berry, C.F. Schmidt, and A.J. Turberfield, *Rapid chiral assembly of rigid DNA building blocks for molecular fabrication*, Science **310** (2005), 1661-1665.

[32] Y. He, T. Ye, M. Su, C. Zhang, A.E. Ribbe, W. Jiang, and C. D. Mao, *Hierarchical self-assembly of DNA into symmetric supramolecular polyhedra*, Nature **452** (2008), 198-201.

[33] N.C. Seeman, *DNA nicks and nodes and nanotechnology*, NanoLett. **1** (2001), 22-26.

[34] T.-J. Fu, Y.-C. Tse-Dinh, and N.C. Seeman, *Holliday junction crossover topology*, J. Mol. Biol. **236** (1994), 91-105.

[35] W. Sun, C. Mao, H. Iwasaki, B. Kemper, and N.C. Seeman, *No braiding of Holliday junctions in positively supercoiled DNA molecules*, J. Mol. Biol. **294** (1999), 683-699.

[36] T.-J. Fu and N.C. Seeman, *DNA double crossover molecules*, Biochem. **32** (1993), 3211-3220.

[37] A. Schwacha and N. Kleckner, *Identification of double Holliday junctions as intermediates in meiotic recombination*, Cell **83** (1995), 783-791.

[38] X. Zhang, H. Yan, Z. Shen, and N.C. Seeman, *Paranemic cohesion of topologically-closed DNA molecules*, J. Am. Chem. Soc. **124** (2002), 12940-12941.

[39] H. Qiu, J.C. Dewan, and N.C. Seeman, *A DNA decamer with a sticky end*, J. Mol. Biol. **267** (1997), 881-898.

[40] M.H. Caruthers, *Gene synthesis machines: DNA chemistry and its uses*, Science **230** (1985), 281-285.

[41] J. Chen and N.C. Seeman *The synthesis from DNA of a molecule with the connectivity of a cube*, Nature **350** (1991), 631-633.

[42] Y. Zhang and N.C. Seeman *A solid-support methodology for the construction of geometrical objects from DNA*, J. Am. Chem. Soc. **114** (1992), 2656-2663.

[43] J. Qi, X. Li, X. Yang, and N.C. Seeman, *The ligation of triangles built from bulged three-arm DNA branched junctions*, J. Am. Chem. Soc. **118** (1996), 6121-6130.

[44] P. Sa-Ardyen, A.V. Vologodskii, and N.C. Seeman, *The flexibility of DNA double crossover molecules*, Biophys. J. **84** (2003), 3829-3837.

[45] N.C. Seeman, J.M. Rosenberg, and A. Rich, *Sequence specific recognition of double helical nucleic acids by proteins*, Proc. Nat. Acad. Sci. (USA) **73** (1976), 804-808.

[46] S.M. Freier and K.-H. Altmann, *The ups and downs of nucleic acid duplex stability: Structure-stability relationships on chemically-modified DNA:RNA duplexes*, Nucl. Acids Res. **25** (1977), 4429-4443.

[47] P. E. Nielsen, M. Egholm, R.H. Berg and O. Buchardt, *Sequence selective recognition of DNA by strand displacement with a thymine-substituted polyamide*, Science **254** (1991), 1497-1500.

[48] P.S. Lukeman, A. Mittal, and N.C. Seeman, *Two-dimensional PNA/DNA arrays: Estimating the helicity of unusual nucleic acid polymers*, Chem. Comm. **2004** (2004), 1694-1695.

[49] J. Chen and N.C. Seeman, *The electrophoretic properties of a DNA cube and its sub-structure catenanes*, Electrophoresis **12** (1991), 607-611.

[50] Y. Zhang and N.C. Seeman, *The construction of a DNA truncated octahedron*, J. Am. Chem. Soc. **116** (1994), 1661-1669.

[51] J.H. White, K.C. Millett, and N.R. Cozzarelli, *Description of the topological entanglement of DNA catenanes and knots by a powerful method involving strand passage and recombination*, J. Mol. Biol. **197** (1987), 585-603.

[52] K. Mizuuchi, B. Kemper, J. Hays, and R.A. Weisberg, *T4 endonuclease VII cleaves Holliday structures*, Cell **29** (1982), 357-365.

[53] H. Yan, X. Zhang, Z. Shen, and N.C. Seeman, *A robust DNA mechanical device controlled by hybridization topology*, Nature **415** (2002), 62-65.

[54] N.C. Seeman, J. Chen, S.M. Du, J.E. Mueller, Y Zhang, T.-J. Fu, Y. Wang, H. Wang, and S. Zhang, *Synthetic DNA knots and catenanes*, New J. Chem. **17** (1993), 739-755.

[55] N.C. Seeman, *The design of single-stranded nucleic acid knots*, Molec. Eng. **2** (1992), 297-307.

[56] S.M. Du and N.C. Seeman, *The construction of a trefoil knot from a DNA branched junction motif*, Biopolymers **34** (1994), 31-37.

[57] J. E. Mueller, S.M. Du, and N.C. Seeman *Design and synthesis of a knot from single-stranded DNA*, J. Am. Chem. Soc. **113** (1991), 6306-6308.

[58] H. Wang, S.M. Du, and N.C. Seeman, *Tight DNA knots*, J. Biomol. Struct. & Dyns. **10** (1993), 853-863.

[59] S. M. Du and N.C. Seeman, *Synthesis of a DNA knot containing both positive and negative nodes*, J. Am. Chem. Soc. **114** (1992), 9652-9655.

[60] S.M. Du, B.D. Stollar, and N.C. Seeman, *A synthetic DNA molecule in three knotted topologies*, J. Am. Chem. Soc. **117** (1995), 1194-1200.

[61] S.M. Du, H. Wang, Y.-C. Tse-Dinh, and N.C. Seeman, *Topological transformations of synthetic DNA knots*, Biochem. **34**, 673-682 (1995).

[62] H. Wang, R.J. Di Gate, and N.C. Seeman, *An RNA topoisomerase*, Proc. Nat. Acad. Sci. (USA) **93** (1996), 9477-9482.

[63] S.M. Du, S. Zhang, and N.C. Seeman, *DNA junctions, antijunctions and mesojunctions*, Biochem. **31** (1992), 10955-10963.

[64] H. Wang and N.C. Seeman, *Structural domains of DNA mesojunctions*, Biochem. **34** (1995), 920-929.

[65] C. Liang and K. Mislow, *On Borromean links*, J. Math. Chem. **16** (1994), 27-35.

[66] P. Cromwell, E. Beltrami, and M. Rampichini, *The Borromean rings*, Math. Intelligencer **20** (1998), 53-62.

[67] C. Mao, W. Sun, and N.C. Seeman, *Assembly of Borromean rings from DNA*, Nature **386** (1997), 137-138.

[68] G. Felsenfeld, D.R. Davies, and A. Rich, *Formation of a three-stranded polynucleotide molecule*, J. Am. Chem. Soc. **79** (1957), 2023-2024.

[69] E. Winfree, F. Liu, L.A. Wenzler, and N.C. Seeman, *Design and self-assembly of two-dimensional DNA crystals*, Nature **394** (1998), 539-544.

[70] P.W.K. Rothemund, N. Papadakis, and E. Winfree, *Algorithmic self-assembly of DNA Sierpinski triangles*, PLOS Biology **2** (2004), 2041-2053.

[71] C. Mao, P.E. Constantinou, F. Liu, Y. Pinto, J. Kopatsch, P.S. Lukeman, T. Wang, B. Ding, H. Yan, J.J. Birktoft, R. Sha, H.Zhong, L. Foley, L.A. Wenzler, R. Sweet, M. Becker, and N.C. Seeman, *The design of self-assembled 3D DNA networks*, Proc. Intl. Symp. on Nanoscale Devices, Materials, and Biological Systems, 206th Meeting of the Electrochemical Society, Honolulu, PV 2004-XX, Editors: M. Cahay, M. Urquidi-Macdonald, S. Bandyopadhyay, P. Guo, H. Hasegawa, N. Koshida, J.P. Leburton, D.J. Lockwood, S. Seal, and A. Stella., vol. 13, 509-520, 2005.

[72] T. LaBean, H. Yan, J. Kopatsch, F. Liu, E. Winfree, J.H. Reif, and N.C. Seeman, *The construction, analysis, ligation and self-assembly of DNA triple crossover complexes*, J. Am. Chem. Soc. **122** (2000), 1848-1860.

[73] F. Mathieu, S. Liao, C. Mao, J. Kopatsch, T. Wang, and N.C. Seeman, *Six-helix bundles designed from DNA*, NanoLett. **5** (2005), 661-665.

[74] P. Constantinou, T. Wang, J. Kopatsch, L.B. Israel, X. Zhang, B. Ding, W.B. Sherman, X. Wang, J. Zheng, R. Sha, and N.C. Seeman, *Double cohesion in structural DNA nanotechnology*, Org. & Biomol. Chem. **4** (2006), 3414-3419.

[75] B. Ding, R. Sha and N.C. Seeman, *Pseudohexagonal 2D DNA crystals from double crossover cohesion*, J. Am. Chem. Soc. **126** (2004), 10230-10231.

[76] J. Zheng, P.E. Constantinou, C. Micheel, A.P. Alivisatos, R.A. Kiehl, and N.C. Seeman, *2D nanoparticle arrays show the organizational power of robust DNA motifs*, NanoLett. **6** (2006), 1502-1504.

[77] R.D. Barish, P.W.K. Rothemund, and E. Winfree, *Two computational primitives for algorithmic self-assembly: copying and counting*, NanoLett. **5** (2005), 2586-2592.

[78] F. Liu, R. Sha, and N.C. Seeman, *Modifying the surface features of two-dimensional DNA crystals*, J. Am. Chem. Soc. **121** (1999), 917-922.

[79] D. Liu, M.S. Wang, Z.X. Deng, R. Walulu, and C.D. Mao, *Tensegrity: construction of rigid DNA triangles with flexible four-arm junctions*, J. Am. Chem. Soc. **126** (2004), 2324-2325.

[80] Z. Shen, H. Yan, T. Wang, and N.C. Seeman, *Paranemic crossover DNA: a generalized Holliday structure with applications in nanotechnology*, J. Am. Chem. Soc. **126** (2004), 1666-1674.

[81] B. Yurke, A. J. Turberfield, A.P. Mills, Jr., F. C. Simmel, and J.L. Newmann, *A DNA-fuelled molecular machine made of DNA*, Nature **406** (2000), 605-608.

[82] S. Liao and N.C. Seeman, *Translation of DNA signals into polymer assembly instructions*, Science **306** (2004), 2072-2074.

[83] H. Zhong and N.C. Seeman, *RNA used to control a DNA rotary machine*, NanoLett. **6** (2006), 2899-2903.

[84] B. Ding and N.C. Seeman, *Operation of a DNA robot arm inserted into a 2D DNA crystalline substrate*, Science **314** (2006), 1583-1585.
[85] H. Urata, K. Shinohara, E. Ogura, Y. Ueda, and M. Akagi, *Mirror-image DNA*, J. Am. Chem. Soc. **113** (1991), 8174-8175.
[86] L. Zhu, P.S. Lukeman, J. Canary, and N.C. Seeman, Nylon/DNA: *Single-stranded DNA with a covalently stitched nylon lining*, J. Am. Chem. Soc. **125** (2003), 10178-10179.
[87] G. Buck, R.G. Scharein, J. Schnick, and J. Simon, *Accessibility and occlusion of biopolymers, ray tracing of radiating tubes, and the temperature of a tangle*, Phys. Rev. E **77** (2008), 011803-011806.
[88] R. Saiki, T.L. Bugawan, G.T. Horn, K.B. Mullis, and H.A. Ehrlich, *The analysis of enzymatically labeled β-globin DQαDNA with allele-specific oligonucleotide probes*, Nature **324** (1986), 163-166.
[89] P.W. K. Rothemund, *Folding DNA to create nanoscale shapes and patterns*, Nature **440** (2006), 297-302.
[90] H. Yan, T. LaBean, L.P. Feng, and J.H. Reif, *Directed nucleation assembly of DNA tile complexes for bar-code patterned lattices*, Proc. Nat. Acad. Sci. (USA) **100** (2003), 8103-8108.
[91] N.C. Seeman, *DNA in a material world*, Nature **421** (2003), 427-431.
[92] N.C. Seeman, *Macromolecular design, nucleic acid junctions and crystal formation*, J. Biomol. Struct. & Dyns. **3** (1985), 11-34.
[93] N.C. Seeman, *From Genes to machines: DNA nanomechanical devices*, Trends Biochem. Sci. **30** (2005, 119-125).
[94] B.H. Robinson and N.C. Seeman *The design of a biochip*, Prot. Eng. **1** (1987), 295-300.
[95] A. Chworos, I. Severcan, A.Y. Koyfman, P. Weinkam, P. Oroudjev, H.G. Hansma, and L. Jaeger, *Building programmable jigsaw puzzles with RNA*, Science **306** (2004), 2068-2072.

Department of Chemistry, New York University, New York, NY 10003, USA
E-mail address: `ned.seeman@nyu.edu`

Proceedings of Symposia in Applied Mathematics
Volume **66**, 2009

Long Tangled Filaments

Jonathan Simon

ABSTRACT. We are interested in filaments, from rope and string and hair to DNA and proteins, anything that might be understood as one-dimensional strands wiggling and tangling in three-dimensional space.

If the filaments are short, we can try to describe the exact geometric shape and understand how the shape relates to physical behavior. If the filaments are somewhat long and flexible, then topological knot type can be measured and used, as evidenced by the success of topological methods for studying the actions of DNA enzymes. But if the filaments are very long (think of a complicated 3-dimensional scribble) or somehow random (think of a statistical ensemble of complicated 3-dimensional scribbles) then it may be impractical to try describe the exact shapes or even knot types. We need to develop a vocabulary of ideas and models that describe physically important geometric/topological properties of long tangled things. In particular, we want to make geometric measurements on a filament that help us understand how tangled it is.

In this paper, we consider some ideas, experiments, and theorems dealing with packing, curvature, tangling, and knotting of individual filaments, starting with simple ones and moving on to complicated ones, as well as "random" collections. We explore some of the work that has been done, some open questions, and include some topics that seem well-suited for undergraduate research activities.

1. Introduction: People

Usually, one starts a paper with the math and ends with a section for acknowledgements. But the whole area of "physical knots" has developed so collaboratively, that it seems more appropriate to begin with people. The first five colleagues listed below are recent research collaborators; the others have been generous with helpful discussions to frame questions or find answers about "tangling".

- Greg Buck
- Eric Rawdon
- Rob Scharein
- Travis Pogmore
- Mark Tucker

2000 *Mathematics Subject Classification.* 57M25, 53A04, 82D60.

Some of the research described here was partially supported by NSF Grant DMS 0107209.

- Javier Arsuaga, Isabel Darcy, Yuanan Diao, Jason Cantarella, Andreas Hanke, Rob Kusner, Steve Levene, Peter Røgen, Heather Russell, Andrzej Stasiak, DeWitt Sumners, (apologies and thanks to other friends and colleagues I am undoubtedly omitting.).

FIGURE 1. Photo of some of the "physical knots" researchers from the workshop on "Mathematics of Knotting and Linking in Polymer Physics and Molecular Biology", held at the Banff International Research Station, Banff, Canada, May 21–25, 2007

2. Three scales of filaments

In "classical" knot theory, the knots are conceived as one-dimensional abstractions, with no imposition of physical properties such as length, thickness, flexibility or elasticity. For example, a standard proof that knot-composition is a commutative operation rests on the ability to shrink one of the factors until it is very small; then deform the string to slide the little knot though the big one to move it to the other side (see Figure 2) .

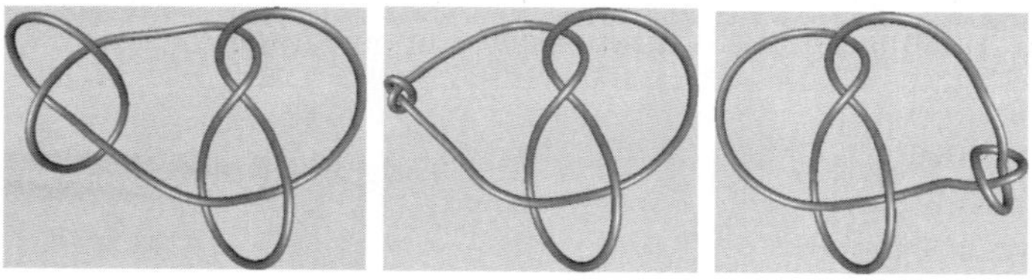

FIGURE 2. Excerpt from typical proof that knot composition is topologically commutative

But if we assume that the filaments are made of some physically real material, or just add some physical-like properties to the mathematical abstractions, then we have to confront geometric limitations on movement. With such constraints, we can identify three fundamentally different size scales for filamentary structures.

2.1. Small scale: Rigid geometry. A polymer can be very rigid, if we consider just a few monomers; or it can be flexible if we consider longer strings. In Figure 3, we see nano-structures built from DNA "sticks". Here a a feature of rigid geometry such as the angle at which lines meet is important. Small rigid assemblies might be used as building-blocks for larger molecular lattices. (See the article by N. Seeman in this volume.)

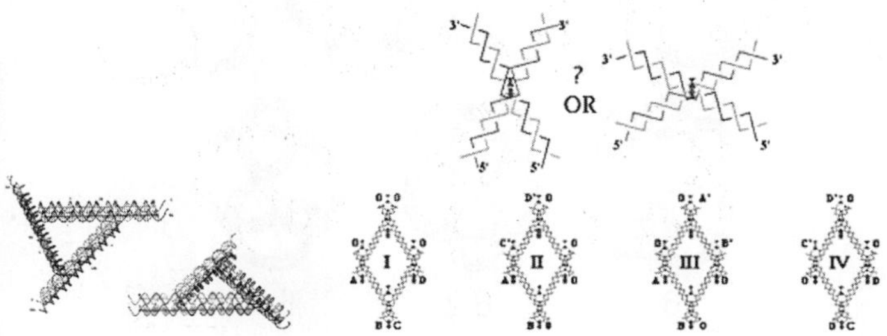

FIGURE 3. Nano-scale structures where filaments are essentially rigid sticks. (Images courtesy of Nadrian C. Seeman; see `http://seemanlab4.chem.nyu.edu/nanotech.html`, and `http://seemanlab4.chem.nyu.edu/Bowtie.html`)

2.2. Medium scale: Topological knot type. In the middle range, one is studying filaments where the combination of length and local flexibility allows tying knots, yet restricted enough that the knots are topologically understandable. This is the scale where researchers developed tables of knot-types (as in Figure 4)

This is the range of size where researchers have had impressive success analyzing the action of DNA enzymes using topology. In a typical experiment with double-stranded DNA, somewhat tangled loops of around 10 kbp size are acted upon by a topoisomerase whose local spatial effect on the DNA plasmid would appear as changing a crossing in a knot diagram. By comparing the [distributions of] initial knots to the [distributions of] resulting knots, molecular biologists can make decisions about how the enzymes are acting.

The approach to enzyme action that is nicknamed "topological enzymology" has also led to fundamental mathematics problems about which knots and links can be obtained from which others under one, or a succession of, crossing-changes or other simple transformations. There are a number of papers that include historical perspective and many references for both the science and the math, as well as their own current research contributions; we refer the reader to papers such as [**Sum95, Sum07, BF07a, BF07b, DSS08**]

Here is the basic situation: Suppose we have some knot ("knot" here might be a single component, or a multi-component link) in 3-space. We isolate part of the knot by surrounding it with a 3-ball and replace the portion of the knot inside the ball with some other ensemble of strings having the same number of endpoints (imagine inserting various tangling patterns of two strands in the ball B

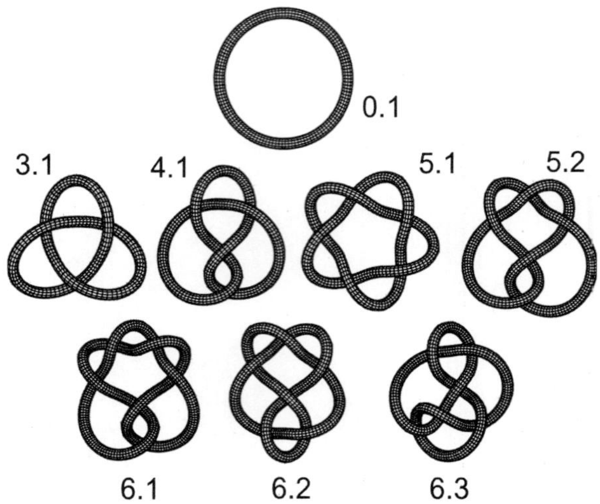

FIGURE 4. Knots organized by topologically minimum crossing number

in Figure 5). One can think in terms of four topological "variables": the original knot, the part that is trapped in the ball, the tangling pattern we replace that with, and the knot that results. Mathematicians have developed theorems relating these to each other. Perhaps the "simplest"case (which is not at all simple) is the question of what can happen if one changes a single crossing in a knot diagram (see for example [**Sch85**] and other literature on knots with "unknotting number one"). By imposing limitations, imposed by chemical feasibility, on some or all of the structures, the topological freedom is reduced. This science of physically constrained knot theory has been used effectively to illuminate, sometimes even replace, laboratory experiments. Meanwhile, knot theorists have been led to new challenging questions.

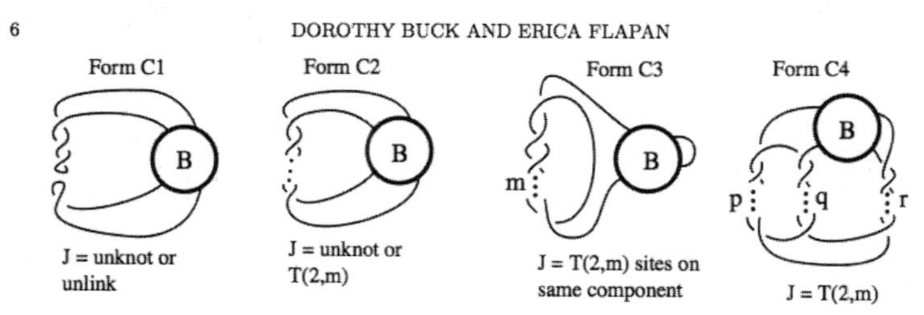

FIGURE 5. Inserting tangles (diagram courtesy of Dorothy Buck and Erica Flapan)

2.3. Large scale: Long and flexible. In this range of size and/or flexibility, the filaments tangle in ways where topological analysis does not seem feasible ("Is this a 100-crossing knot type or a 99-crossing knot type?"), and might even miss the essential nature of the structure (think of highly tangled unknots). We will describe three examples in this size range where it seems important to analyze tangling as something other than rigid geometry or topological knot/link type. In later sections, we will develop some of the mathematics more carefully.

2.3.1. *Protein shape descriptors.* In Figure 6 we see a protein molecule in a crystallized conformation, taken from the RCSB Protein Data Bank. Using a java application written by student M. Tucker (and with thanks to P. Røgen for a helpful earlier conversation), we downloaded this structure, stripped it down to the α-carbon backbone, and imported that into KnotPlot. The "real" data at this point is just the $\mathbb{R}^3$ coordinates of 257 atoms. The (backbone) molecular bond-graph can be viewed as a polygonal arc with 257 vertices. It is easier to see the overall shape and tangling if we smooth and thicken the arc, which is done with KnotPlot's viewing options. Since the filament is an arc with free ends, it is topologically unknotted, which we can see by gradually relaxing the structure as shown in Figure 6.

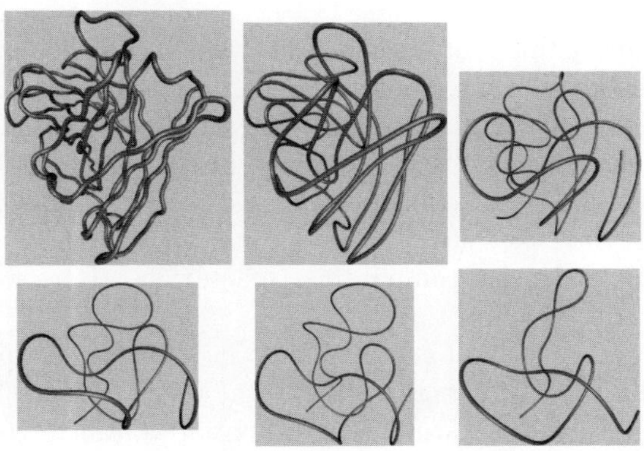

FIGURE 6. Protein (α-carbon backbone) with important tangling but trivial topology revealed by relaxing

In papers such as [**Art97, RB03, RF03**] , the tangling of proteins is measured using the average crossing number (see Section 5 for discussion of ACN) and related indices; these quantities either measure directly, or correlate in provocative ways with, biochemical properties.

2.3.2. *Spatial packing regimes.* (This "universal" view of packing and tangling is a conceptualization that Greg Buck has patiently taught the author over a number of years.)

The tangling in a polymer system is intimately connected to its spatial packing density. The spatial packing (as some kind of average over the given system) depends on many factors such as polymer length and bond-rigidity, relative amounts of polymer vs solvent, detailed chemistry of the polymer/solvent interactions in the given system, electrostatics within the polymer, and temperature. We believe that the connection between packing density and tangling can best be understood by focusing on three basic packing "regimes".

In [**BS07a**] we distinguish three basic spatial packing regimes for filaments (which we might think of as 1-dimensional cores of structures with thickness). The taxonomy is by average spatial density of filament length. If the packing is "random" or "generic" then there is a predicted amount of tangling, as measured by average crossing number (again, see see Section 5 for details ACN). We assert that when the tangling is different from the predicted rate, that is evidence of some control or other special feature of the process that is generating the filaments. To put this another way: In a given process that generates filaments, there is an average spatial density and a resulting expected average amount of ACN. If the ACN is different from the expected amount, there is some special feature of the system that is causing this and that should be important to understand.

Here is a short description of the packing regimes; we do somewhat more careful math later (Section 7). See Figure 7 to visualize these three fundamental kinds of packing. For the middle regime, which seems the most subtle, we have two pictures in Figure 7: the first is a neat mat where the tangling is much higher than we would expect for random packing at this density; the second is an equilateral N-edge polygon generated by a random process, and for such polygons, the ACN is of the expected order $N \log N$ ([**DDK$^+$03**]).

Linear: Around any point x of the filament, the total length of filament that lies within distance ρ of x is (on average, approximately) proportional to ρ. Here we expect the ACN to grow linearly with length L.

Laminar: Around any point x of the filament, the total length of filament that lies within distance ρ of x is proportional to ρ^2. If the filament has some thickness, then the "stuff" fills a virtual surface or has a spatially fractal-like shape of density comparable to a surface. Here we expect the ACN to grow in proportion to $L\, log\, L$.

Globular: Around any point x of the filament, the total length of filament that lies within distance ρ of x is proportional to ρ^3. If the filament has some thickness, then the "stuff" fills a spatial volume. Here we expect the ACN to grow in proportion to $L^{4/3}$.

If the filament is idealized to have no thickness, then there is a fourth regime: The density grows without bound as we pack more and more length into a finite volume of space. And here we expect the ACN to grow in proportion to L^2.

It is possible to design "by hand"" filament systems where the packing and growth rates (of ACN with length) lie at intermediate levels. But these three (or four) are the fundamental regimes.

The linear and globular regimes seem familiar; but the laminar regime also is believed to occur in important situations. We call it "laminar" because of the packing density; but the interesting physical instances involve filaments that wander around in 3-dimensional space. The paper [**LZC02**] proposes this kind of packing for proteins at medium densities. The picture in Figure 8 is a (very abstract) model we might make of their proposed packing. Imagine a filament that is packed so as to occupy nearly all the space represented by the surfaces of a bunch of packed spheres (not occupying much of the interiors). The left picture shows the (virtual) spheres. The right picture shows a 1-skeleton of those surfaces. Imagine a single filament

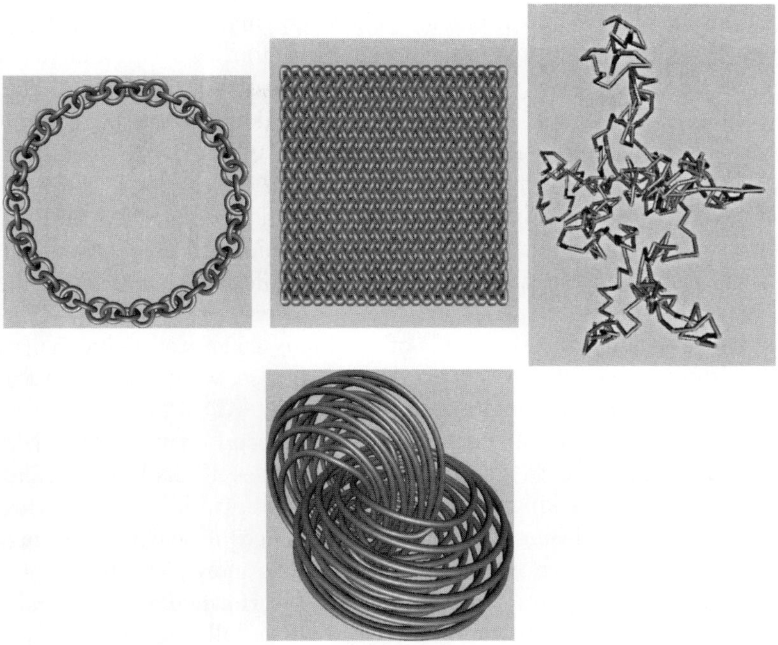

FIGURE 7. Linear, Laminar (non-random w high ACN), Laminar (random with expected ACN), and Globular packing

that wanders around in space so it looks approximately like that 1-skeleton. *What is the tangling complexity of that filament?*

More generally, mathematicians and scientists have analyzed theoretical and computational models of polymers (in particular [**DDK$^+$03, DE05**]) and found the $N \log(N)$ behavior. However, there is still some uncertainty about whether the growth rate is logarithmic or follows some fractional power law if the models are endowed with stronger self-avoidance than what occurs "naturally" in 3-dimensional random walks (see e.g. [**DH08**]).

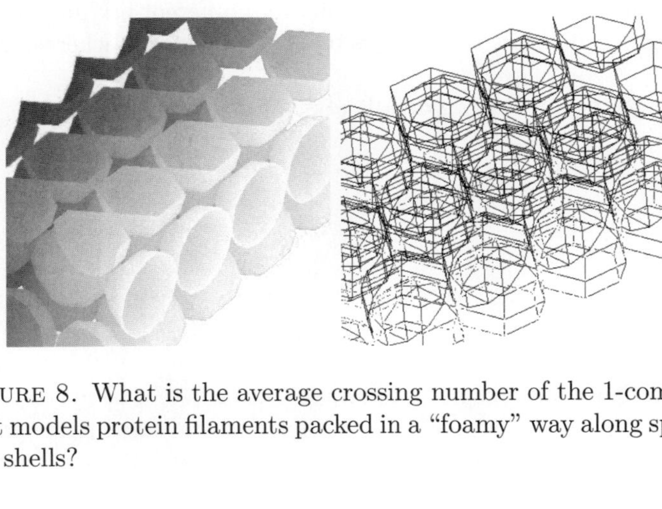

FIGURE 8. What is the average crossing number of the 1-complex that models protein filaments packed in a "foamy" way along spherical shells?

2.3.3. *Different kinds of "random"*. There are many ways to generate "random" filaments. One can, for example, consider random walks on a lattice, random equilateral or non-equilateral polygonal walks without a lattice, or random closed polygons obtained by several different ways of shaking a starting configuration. The tangling, as measured by ACN (or the ratio of ACN to other possible tangling measures), can be used to distinguish different kinds of "random". From the point of view of trying to accurately model polymers and studying the resulting ACN, this is being well-studied (e.g. the references in Section 2.3.2). But from the point of view of creating ways to distinguish between different ideas of "geometrically random", this seems an interesting new area for research projects.

In Figure 9, we illustrate two kinds of randomly generated polygons (produced by KnotPlot's functions *jitter* and *mangle*. In each case, we started with a 100-edge regular planar polygon. For the polygon on the right, KnotPlot moved each vertex by choosing a direction randomly and a distance uniformly between 0 and M, where M is [very] large compared to the original coordinates. (This looks similar, but is not quite equivalent, to picking 100 points with a uniform distribution throughout a ball, since the unbiased length distribution produces a heavier concentration of points near the origin, which is why the figure looks more densely packed around the center.) For the polygon on the left, KnotPlot performed a large number of so-called "pivots" or "crank-shaft" moves: Do all of the following randomly: Choose two vertices v, w; these cut the polygon into two arcs, A_1, A_2; choose one of the two arcs A_i; choose an angle θ; using the line between v and w as an axis, rotate the arc A_i about the axis through angle θ. This is the same method as used in many polymer simulations.

The two polygons in Figure 9 are made of 100 segments, but randomized differently; One is polymer-like wiggling, with segment lengths fixed; the other is created by a random zig-zag in space which, in particular, produces randomly varying segment lenghts. The two polygons look different, and they have different geometric properties. They have different total lengths, different amounts of bending (see 3.2 below), and different ACNs.

We know that ACN is partly a function of length: If one wiggly filament is longer than another, it will tend to have more self-crossings. To compensate for the length effect of one kind of randomizing versus the the other, we can use more sticks for the crankshaft-randomized polygon. This gives Figure 10. Here the two polygons have the same geometric length but very different packing and ACNs. For the tighter looking filament, the ACN is on the order of L^2, whereas it is $L\log(L)$ for the other.

See Section 8 for a few more details about this example.

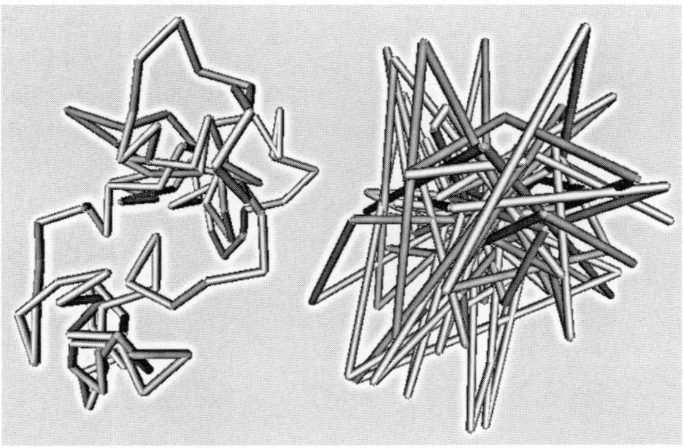

FIGURE 9. Two kinds of "random" 100-edge polygons

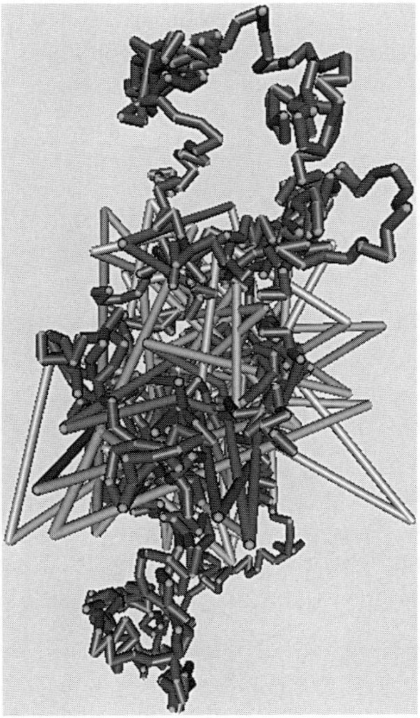

FIGURE 10. Two kinds of "random" polygons: different numbers of sticks to equalize the lengths

3. Geometric properties of a filament

We want to develop some of the geometric properties one can use to describe a filament. It seems easiest to develop the ideas by working with simple shapes, and then go on to apply them to more complicated shapes.

3.1. Geometric properties: Length and thickness. Perhaps the most basic property of a piece of string, K, is its length; we denote that L or $L(K)$. But length alone does not control how much tangling the string can achieve. We can tie many more knots in a one-foot pice of thin thread than in a one-foot piece of thick rope (see Figure 11). Or we could use thick rope, but just have a lot of length. We need to consider the thickness of the filament and its relation to the length.

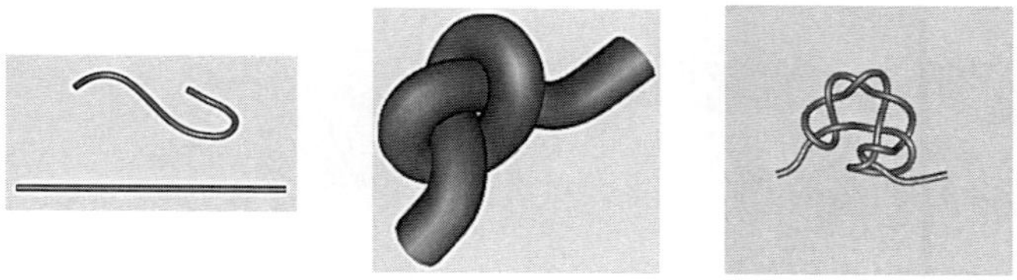

FIGURE 11. Length and thickness

If the filament is essentially a smooth curve, then we can model it as a tube with uniform circular cross-section of radius R, as in Figure 12. (Note: Other authors or manufacturers of rope and fishing line may use the diameter instead of the radius). We define the *Ropelength* of the "rope" $= \frac{L}{R}$. In proving theorems, it often is convenient to normalize a curve so the thickness radius $= 1$, in which case the ropelength is just the length $L(K)$.

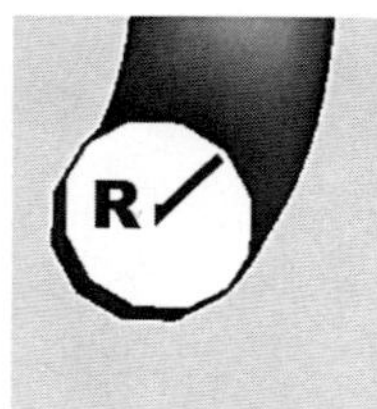

FIGURE 12. Defining "thickness"

This approach to defining thickness of smooth curves was used in [**LSDR99**]. If the curve is actually a polygon, then some care is needed to properly define a "thickness". This was done by E. Rawdon in [**Raw98, Raw00**]. A unified approach that handles both smooth curves and polygons was developed in [**GM99**].

3.2. Geometric properties: Curvature (polygonal and smooth). If K is a polygon, we can measure the turning at each vertex: Walk along the edges (Figure 13), measure the exterior turning angle θ_i at each vertex, and add these angles to get the *total curvature* of the polygon, $\sum_i \theta_i$, which we denote $\kappa(K)$. If the polygon lies in a plane, then we can distinguish positive and negative turnings, and get a signed total curvature; but if K is in 3-space, then we just add the magnitudes of the angles and record a positive $\kappa(K)$.

If the curve is smooth, then we measure how the unit tangent vector $\mathbf{T}$ turns with respect to arclength along the curve. The derivative $\frac{d\mathbf{T}}{ds}$ at each point is a vector normal to the curve, whose length is the curvature. That is

$$\frac{d\mathbf{T}}{ds} = \kappa \mathbf{N}$$

and the total (absolute) curvature is

$$\kappa(K) = \int_K \| \frac{d\mathbf{T}}{ds} \| \, .$$

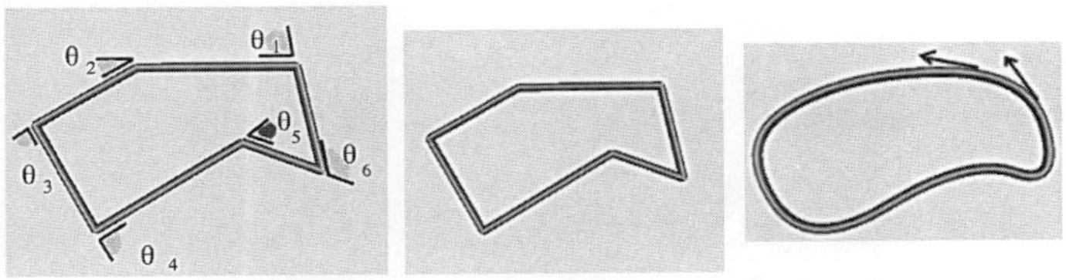

FIGURE 13. Walk along counterclockwise and add exterior angles

From [**Mil50**] we know that the total curvature of a smooth curve is the supremum of the total curvatures of inscribed polygons. We will say more about how curvature does or does not limit tangling in Section 4.4.

3.3. Geometric properties: Sticks. Polymers sometimes are modeled as some number of rigid sticks joined end-to-end; the model might allow free bending at the joints, or impose a penalty based on bending.

FIGURE 14. "Oribasius", 10 ft. garden sculpture, Birmingham, MI, M. Goldman

For polymers, one talks about "persistence length" or "Kuhn statistical segment" length. Imagine walking along a polymer strand: Initially, the geometry of how one monomer is attached to the next gives us no choice about where the next step should aim; but eventually local flexibilities add up to where there is essentially no correlation between the initial position or direction and our latest one (except for some kind excluded-volume interaction that would apply equally well to a piece of the polymer that is completely disconnected from this one) . Here is a more careful description borrowed from `http://en.wikipedia.org/wiki/Persistence_length`

> The persistence length is a basic mechanical property quantifying the stiffness of a long polymer. Informally, for pieces of the polymer that are shorter than the persistence length, the molecule behaves rather like a flexible elastic rod, while for pieces of the polymer that are much longer than the persistence length, the properties can only be described statistically, like a three-dimensional random walk. Formally, the persistence length is

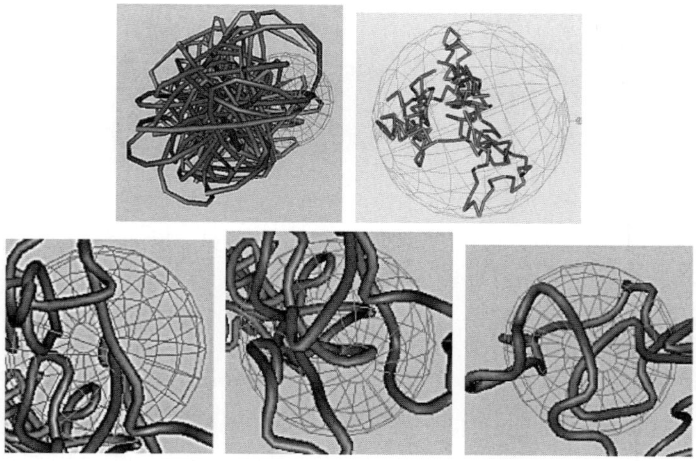

FIGURE 15. Measure average density of length per unit of space-volume over the whole object

defined as the length over which correlations in the direction of the tangent are lost. Let us define the angle θ between a vector that is tangent to the polymer at position 0 (zero) and a tangent vector at a distance L away from position 0. It can be shown that the expectation value of the cosine of the angle falls off exponentially with distance, $\langle cos\,\theta \rangle = e^{-(L/P)}$, where P is the persistence length and $\langle \rangle$ denotes the average over all starting positions.

A piece of cooked spaghetti has a persistence length on the order of 10 cm. Double-helical DNA has a persistence length of about 50 nanometers.

In polymer science, persistence length is one half of the Kuhn length, the length of hypothetical segments that the chain can be considered as freely joined.

Much of "classical" knot theory was developed in the realm of piecewise linear objects and functions. But in that category, subdivision is always allowed and often needed to prove one thing equivalent to another. As soon as we disallow subdivision, that is we restrict to polygons with a fixed number of edges, new mathematical questions arise. We discuss these a little more in Section 4.1.

3.4. Geometric properties: Compaction. Measure the amount of filament length contained in a given volume. This is what we mean by "packing density". (See Figure 15.) For each point $x \in K$ and each radius ρ, let $L(x, \rho)$ denote the amount of filament length contained in a ball of radius ρ centered at point x. To prove theorems relating packing, curvature, and tangling, we will assume that the packing density is homogeneous, that is the function $L(x, \rho)$ is essentially independent of x.

3.5. Geometric properties: Summary.

- Length L
- Thickness radius R snd *ropelength* $\frac{L}{R}$

- Total curvature $\kappa(K)$
- Number of sticks, N
- Compaction $L(x, \rho)$

4. Interaction of geometry and topology

Now we begin to consider questions about how geometric constraints can influence topological complexity.

4.1. Geometry limits topology: Questions about sticks. The first geometric property, number of "sticks", may get a little far from polymer questions, but it is an area where many accessible yet challenging problems can be identified (so a good source of undergraduate research topics). Is is possible to frame questions that students can understand and explore, though definitive theorems may be very difficult and/or deep.

We can specify the kinds of "edges" allowed in a polygon in many ways, for example:

- Straight segments of identical lengths (*equilateral edge number*)
- Straight segments of varying lengths (*edge number*)
- Straight segments of equal, or varying, lengths that meet always at a fixed angle (*equi-angular edge number*) as in Figure 14
- Edges in the 3-dimensional integer lattice (*lattice edge number*)
- Straight segments that are unions of lattice edges (*latice stick number*)

and then ask questions such as

- What is the smallest number of edges needed to make a nontrivial knot?
- What is the minimum edge-number of the figure-eight knot? of 8_{19}?
- How does minimum edge-number behave under knot sum?
- How are the different edge-numbers related?

There is a lot of literature [**Var**] on various kinds of edge-numbers and how they limit topological complexity. We might identify the papers [**Neg91, Dia93, Ran94a, Ran94b**] among the seminal ones for these kinds of questions. This is an area with much work done and many questions open; we will defer to others for more detailed discussion of this topic.

4.2. Geometry limits topology: (Digression ...) Questions about other shapes besides sticks. What if we build knots out of pieces of restricted shape. We can ask questions similar to those for sticks. For example, try to make knots that are "circular polygons" built from quarter-circles (as the common toy shown in Figure 16) or more general arcs of circles.

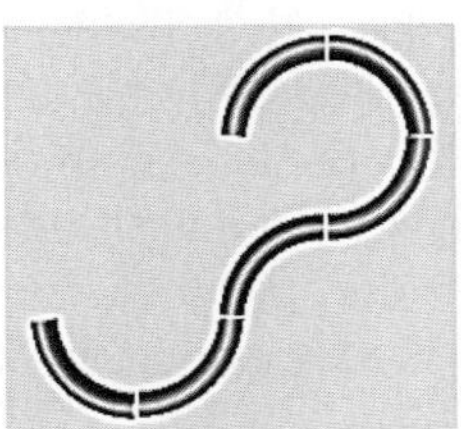

FIGURE 16. Make a "polygon" out of quarter-circles or more general arcs of circles

After a little experimenting, it is easy to get convinced that one cannot build a planar smooth closed curve out of six quarter-circles. But it is a little harder to prove it! (Again, something students can enjoy discovering.)

The question of building knots out of arcs of circles may seem frivolous, but it connects to other problems, e.g [**RST, MG07**].

In [**RST**], we consider the effect on knotted polygons of Möbius transformations of $\mathbb{R}^3$. The operation of takes straight lines to straight lines and/or circles. Since inversion in a round 2-sphere is a homeomorphism of $\mathbb{R}^3$ (extended to include the point at infinity), if we apply a Möbius transformation to a polygonal knot K, we get in general a piecewise-circular "polygon"; of the same knot type as K or its mirror image. So the circle-arc number of a knot type is at most the edge number of the knot type. But probably the circle arc number is smaller.

The point of the paper [**RST**] is not to study the resulting circle-arc figure, but rather to study what happens if we re-connect the inverted vertices of K with straight segments. The knot type can be completely different from K, but the paper shows that only a limited number of knot types can be obtained this way from a given starting polygon.

It is clear that every knot type can be achieved by a piecewise-circular "polygon"; and (as suggested by Figure 16) the curve can be C^1 differentiable; but the normal vectors will have jump-discontinuities. In [**MG07**], it is shown that each knot type can be realized as a C^2 smooth curve of constant (scalar) curvature, using a small library of arcs as building blocks. The paper [**Gho07**] shows, in fact, that C^∞ knots of constant curvature are dense, in a strong way, among all knots; and that topological isotopy implies constant-curvature isotopy for such knots.

4.3. Geometry limits topology: Ropelength. The basic question has been: What is the {shortest, thickest} piece of "rope" that can form a closed knot? i.e.

$$K \text{ nontrivial smooth knot} \implies \frac{L}{R} \geq \;??$$

The first results were obtained [**LSDR99**] by understanding ropelength in terms of local curvature and "critical self-distance" of a curve, and approximating a thick smooth curve with a polygon having few edges.

In Figure 17, we see how several theorems in turn force a closed curve that is knotted to have greater and greater length, ending with our theorem that still forces only about half the length that experimentally seems needed for a nontrivial knot. In Figure 17, (A) shows the length needed, by Fenchel's theorem, to have a closed curve at all; (B) shows the length needed by the Fary-Milnor theorem that a knot has total curvature $> 4\pi$; (C) shows the length needed to avoid being approximated by a 5-edge polygon; and (D) shows the length needed in experiments to actually tie a nontrivial knot. So at that point, the theory was able to account for only 50% of the experimental value which is approximately 32.74 [**Buc, KBM$^+$96, Pie98, LSDR99, PP01, PP02, MR03, CLMS05, BPR05**].

Over several years, other researchers have proved better and better theorems. (Thanks to Y. Diao for providing the following summary.)

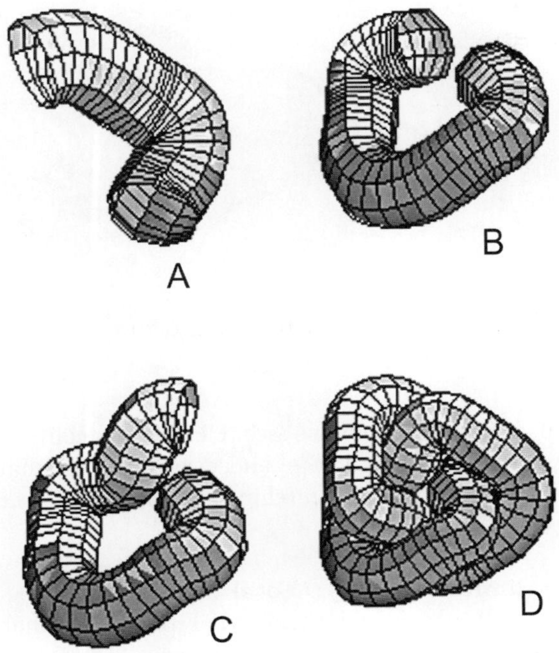

FIGURE 17. Trying to tie a knot with "thick rope"

Authors	Year	L/R is
Litherland *et al*	1999	$> 5\pi \approx 15.71$
Cantarella *et al*	2002	$\geq 2(2+\sqrt{2})\pi \approx 21.45$
Diao	2003	> 24
Denne, Diao, Sullivan	2005	> 31.32
Experiments	1990-2004	≈ 32.74

It may be that more insights and machinery can be developed to push the theory further, though there is no reason to think this would be easy. Meanwhile, the question of finding theoretical lower bounds for ropelength of knot types other than the trefoil (except for likely low estimates from total curvature or edge-number bounds such as in [**Mil50, LSDR99**]), is open and seems more approachable.

We can list some challenges:

(1) Establish a close theoretical bound on how much rope it takes to tie a figure-eight knot, or any others beyond the trefoil.
(2) Is the ropelength minimizer of a knot type $[K]$ unique? Are there multiple minima? multiple local minima?

For single-component knots, conjecture: Within each knot type $[K]$, there is an unique absolute minimizer for ropelength; but there may be different local minimizers. It is easy to construct (thickened) composite knots where it appears impossible to deform one to the (topologically equivalent) other without making the knots thinner or longer. Think, for example, of thickened knots (see Figure 18) $3_1\#4_1\#3_1\#4_1$ and its topological equivalent $3_1\#3_1\#4_1\#4_1$.

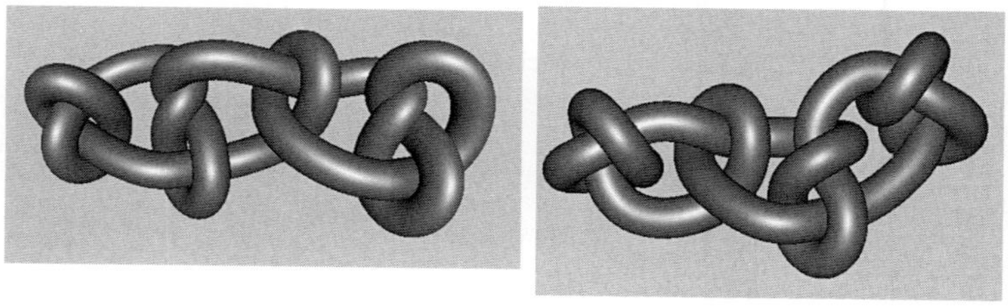

FIGURE 18. Can one thick knot be changed to the other without making the rope thinner or longer?

It seems intuitively plausible (see [**PPS01**] for computer experiments) that there are thick prime knots, and even unknots, that cannot be untangled without thinning or stretching the tube. BUT there are so far no theorems about this.

4.4. Geometry limits topology: Total curvature. If the total curvature of a filament is very small, it cannot be knotted. From a combination of [**Hop35, Whi37, Fen51, Fár49, Mil50**], we have the following "classical" theorems.

	Polygon	Smooth Curve		
Planar simple closed curve	$\sum_i \theta_i$	$\int_K \kappa(x)\, ds$	$=$	2π
Planar and non-convex	$\sum_i \lvert\theta_i\rvert$	$\int_K \lvert\kappa(x)\rvert\, ds$	$>$	2π
Non-planar	$\sum_i \lvert\theta_i\rvert$	$\int_K \lvert\kappa(x)\rvert\, ds$	$>$	2π
Non-trivial knot	$\sum_i \lvert\theta_i\rvert$	$\int_K \lvert\kappa(x)\rvert\, ds$	$>$	4π

So a nontrivial knot has total curvature $> 4\pi$. Can we say more? The theorem in [**Mil50**] actually gives

$$2\pi\, \beta([K]) < \kappa(K)$$

where $\beta[K]$ is Schubert's *bridge number* ([**Sch54**] of the knot type.

But within a fixed amount of total curvature (e.g. arbitrarily close to 4π) we can realize infinitely many knots with arbitrarily high crossing number (see Figure 19). Total curvature can be used to bound topological complexity, but only if we combine it with some other geometric property of the curve, such as thickness: We have shown that for curves of a given ropelength, the number of crossings in a knot cannot grow faster than the total curvature ([**BS07b**] (more on this later).

4.5. Geometry can encourage complexity: Packing. There is at least one way that geometric constraints can *encourage* a filament to look complex (which is not the same as forcing tangling complexity, but might lead to that if the system had some randomness): if we have a long filament packed in a small ball, then it has to curve a lot. This result, in different versions, is attributed in [**Tab03**] to G. Chakerian [**Cha64**] and to I. Fáry [**Fár50**]. A lot of intuitive insights and several different versions and proofs of the theorem are given in [**Tab03**]. The proof below is based on Chakerian's, which also is included in [**Tab03**].

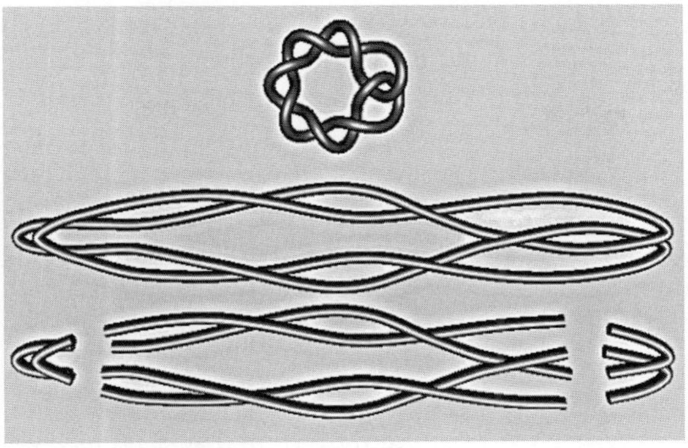

FIGURE 19. Many crossings are possible despite the bound on total curvature

THEOREM 4.1. *Suppose K is a closed curve (think of this as being in $\mathbb{R}^2$ or $\mathbb{R}^3$) having total length L. If K is contained in a round ball of radius ρ, and $L > 2\rho$, then*

$$\textit{total curvature of } K \geq \frac{L}{\rho}.$$

PROOF. (We give a proof assuming K is a smooth closed curve; one can modify the argument to handle piecewise-smooth curves, and modify the statement to include the case of a non-closed curve.) (Note also that the hypothesis $L > 2\rho$ is necessary: until that size, K might just be a straight segment.)

First rescale the picture so the ball has radius $= 1$; our goal is now to show $\kappa(K) \geq L$. Let $s \to x(s)$ be a unit-speed parameterization of K, where $x(0)$ is some point of K and $x(L) = x(0)$. Translate axes so the center of the ball is the origin. The velocity vector $x'(t)$ is a unit vector, and each position $x(t)$ has $|x(t)| \leq 1$. Since $|x'| = 1$, we have the dot product $(x' \cdot x') = 1$, and so the identity $L = \int_{s=0}^{L} (x' \cdot x')\, ds$. Now rewrite that integral using integration by parts:

$$L = \int_{s=0}^{L} (x' \cdot x')\, ds = (x' \cdot x) \Big|_{s=0}^{s=L} - \int_{s=0}^{L} (x \cdot x'')\, ds\,.$$

Since the curve is smooth at the point $x(0) = x(L)$, we have $x(0) = x(L)$ and $x'(0) = x'(L)$, so the first term $= 0$. Thus

$$L \leq \left| \int_{s=0}^{L} (x \cdot x'')\, ds \right| \leq \int_{s=0}^{L} |x|\, |x''|\, ds \leq \int_{s=0}^{L} |x''|)\, ds = \kappa(K)\,. \qquad \square$$

5. Average crossing number

The fundamental problem we face is how to define and measure “tangling”. In this essay, we concentrate primarily on the idea of *average crossing number*, nickname ***ACN***. Roughly: look at a filament and count how many times it crosses over itself; now look from every direction and take the average. At first glance, this is a measure of compaction and self-occlusion or opacity ([**Gra01**]), rather than “tangling”. But our own computer experiments (see Section 9) with randomly generated filaments, and the intuition of scientists who have been using average crossing number to quantify tangling, suggest that ACN is a valid, if indirect, measure of what we really mean by "tangling". We are exploring another more intuitively direct way to define and measure tangling [**SST**], called *average bridging number*: So far, we have found that in random systems, ACN correlates well with this other measure (helping to validate the use of ACN to quantify average tangling over systems of many filaments). At the same time, those correlations vary in specific ways, according to the kinds of filament systems (e.g different notions of "random"), suggesting that this new approach will provide insights beyond ACN.

In order to further develop our understanding of how geometric properties of a filament can limit its tangling or topological complexity, we want to get beyond the question, "Can one make a nontrivial knot if the rope is ...?". Let us allow the possibility that the exact topology of the filament is unknown (maybe so complicated as to be in a practical sense unknowable) and just focus on trying to quantify how much the filament is tangled with itself, using the notion of average crossing number to try to do this. In contrast to topological knot-type, the idea of average crossing number gives us a way to quantify the amount of "tangling" in cases where

- the filament is too long and complicated for topological knot type to be practical, or
- the filament has free ends, so it is topologically unknotted

In a given curve, the "ACN" counts how often the curve crosses over itself, but ignores the order along the curve in which these crossings occur. So one can argue that the quantity being measured really is not "tangling", but something different (opacity or occlusion) [**Gra01**]. A complicated tangle such as the first shown in Figure 20 and a really untangled curve such as the second shown in Figure 20 can have essentially the same ACN. But it takes special effort to construct untangled complicated looking curves with high ACN. Computer experiments indicate (see later) that ACN is an effective indirect measure of whatever one might call "tangling", especially in systems of "random" filaments (see also comments in [**BSSS08**]); and ACN *is* used and defended in polymer/protein/DNA literature as a tangling descriptor [**EEA00, Art97, AT00, JvRSWW92**], as well as a correlate of other physical properties [**KBM+96, VCL+98, RAT02**].

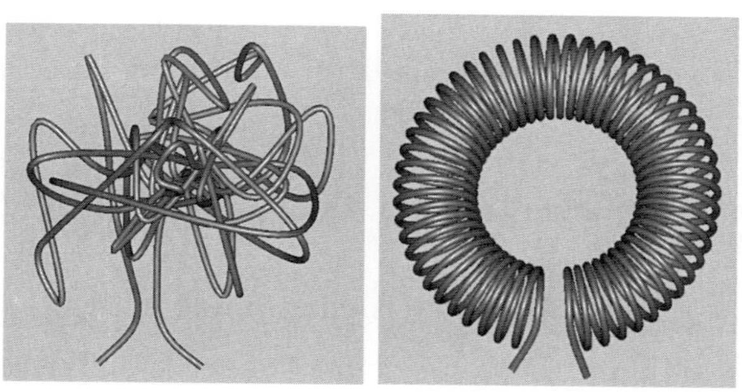

FIGURE 20. Curves with similar *average crossing number* ≈ 470

After we develop the mathematics of ACN more carefully, we will return to the theme of "geometry limits tangling complexity" by describing theorems that give bounds for ACN in terms of ropelength and curvature.

5.1. First definition of *average crossing number*. Count the crossings, but consider "all" directions of looking

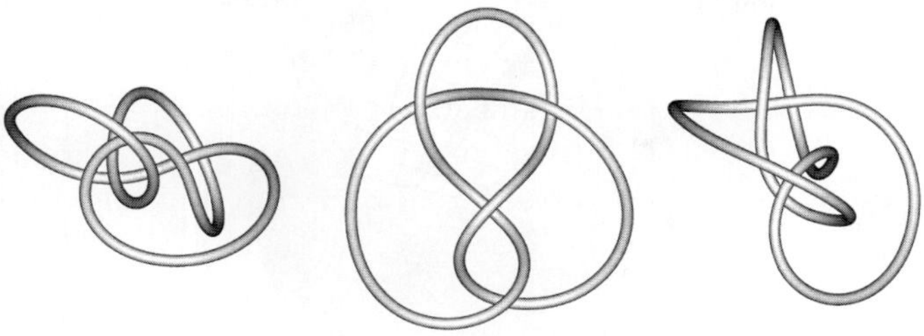

FIGURE 21. Count crossings as seen from different directions, then average

The average crossing number of the space-curve shown in Figure 21 is ≈ 6.3. (This curve happens to be a "figure-eight" knot, whose minimum crossing number is 4, which is what one sees from some fraction of the viewpoints.)

Of course, one cannot really look from "all" directions and make a count for each; but we can approximate. The software system KnotPlot, for example, has a built-in function for ACN in which one specifies some number of randomly chosen directions from which to count crossings. The error bound estimate is

$$\frac{ACN}{\sqrt{Nviews}} .$$

So an estimate with 100 views can be trusted to within around 10%, while 10,000 views gives 1% accuracy.

5.2. More careful definition and understanding of *average crossing number*. For each unit vector $\mathbf{u} \in S^2$, consider those pairs of point $x, y \in K$ such that $\mathbf{xy}$ lines up with $\mathbf{u}$. (See Figure 22) If the curve K is a polygon, then we can visualize the 2-sphere as being (multi-)covered by a patchwork of spherical quadrilaterals, where each quadrilateral is the set of points $\mathbf{u}$ that are hit by crossings of a particular pair of edges of the polygon. Note that for each such quadrilateral, there also is the mirror image in the opposite side of the sphere: if $\mathbf{xy}$ lines up with $\mathbf{u}$ then $\mathbf{yx}$ lines up with $-\mathbf{u}$ Define $Cr(K, \mathbf{u})$ = number of crossings seen from direction $\mathbf{u}$.

$$\text{ACN(K)} = \frac{1}{4\pi} \int_{\mathbf{u} \in S^2} Cr(K, \mathbf{u})\, d(\text{area} S^2) .$$

This integral version of ACN is elegant, but not really useful for calculations. However, it leads us to another integral that is both illuminating and useful. For each pair $(x, y) \in K \times K$, there are those points $\mathbf{u} \in S^2$ such that $(x - y)$ lines up with $\mathbf{u}$.

Define a parameterization (with multiplicities) of S^2 by

$$\phi : K \times K \to S^2 \quad (x, y) \to \frac{(x - y)}{\|(x - y)\|} .$$

We can use this parameterization to compute a weighted area of S^2, where the multiplicities correspond exactly to how many crossings each direction sees. This gives us

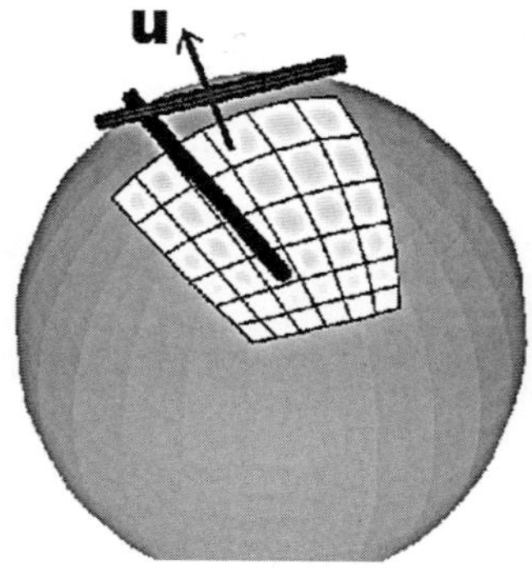

FIGURE 22. Defining ACN

$$\text{ACN(K)} = \frac{1}{4\pi} \int_{(x,y)\in K\times K} |J\phi| \; dxdy \; .$$

where $|J\phi|$ is the Jacobian "stretching factor" for the mapping ϕ.

This formula for ACN was presented in [**FHW94**], and is a variation of Gauss's formula for the linking number of two closed curves in 3-space (see Section 5.2.1 below). The Jacobian can be written as

$$|J\phi| = \frac{\left|(T_x \times T_y) \cdot \frac{(x-y)}{|(x-y)|}\right|}{|x-y|^2} \; .$$

Often the formula is stated with $|x-y|^3$ in the denominator; but we find it helpful to state it with $|x-y|^2$ in the denominator, and the other factor $1/|x-y|$ in the numerator, to remind us that the numerator is just a triple scalar product of three unit vectors.

Here is one intuitive way to understand the formula for the Jacobian: First map $K \times K$ to a singular torus in $\mathbb{R}^3$ by

$$(x, y) \rightarrow x - y \; .$$

The (normal) vector area element on this surface is $(T_x \times T_y)\; dxdy$. Then watch what happens to this surface patch when we project to S^2 in the direction $x - y$: we correct for the tilt of the surface patch relative to the direction of projection by taking the dot product with the unit vector $\frac{(x-y)}{|(x-y)|}$; and, since we are rescaling lengths by $1/|x-y|$, we rescale area by the square of that, hence the $1/|x-y|^2$ in the denominator.

5.2.1. *Variations on the ACN integral.* If x and y come from two different closed curves, and we remove the absolute value from the ACN integrand numerator, we get exactly Gauss's double-integral for the algebraic linking number of closed curves in 3-space. If x, y range over the same curve, but we again remove the absolute value in the numerator of $J\phi$, we obtain the *writhe* of the space-curve, the average [over all directions] of the *signed* crossing number.

6. Combining ACN and other geometric properties to limit or compare tangling

We can bound the average crossing number of a filament in terms of various combinations of ropelength, curvature, and packing [**BS99, BS07b**]. Some bounds are, in a sense, sharp. And for these we assert [**BS07a**] that in random systems, the ACN behavior should look like the bounds. For example, in a particular physical system, if the rate of growth of ACN with ropelength is lower than the predicted bound, we suggest this is a measure of non-randomness in the system. We list the basic theorems below, and indicate which are sharp and where there is a clear need for improvement.

We have various theorems saying that

$$ACN(K) \leq \text{functions}(L, R, \kappa, L(x, \rho)) .$$

Various of the coefficients $\alpha, \beta, \gamma, \delta$ in the list below have been, or could be, improved [**DE02, RS02**] from out original proof for ACN (and a completely different approach for the related *average writhe* [**CDG01**]). Our emphasis right now is on the exponents, so we will leave the coefficients unspecified; but they are specified in [**BS99, BS07b**], and are not silly large numbers. For example, $\beta \approx 0.4$.

- $ACN(K) \leq \alpha \cdot (\frac{L}{R})^2$
  In view of the 4/3 bound below, the exponent "2" cannot be best-possible. However, the coefficient α is smaller than β, so for knots with small ropelength, the quadratic bound can be lower than the 4/3 power bound.
- $ACN(K) \leq \beta \cdot (\frac{L}{R})^{4/3} \quad (\beta > \alpha)$
  The 4/3 power bound is sharp; that is, there exist infinite families of knots and links ([**Buc98, CKS98**]) where the topological minimum crossing numbers grow on the order of the 4/3 power of the ropelengths of the particular curves. So the ACNs of those curves must be growing at least that fast; and our theorem says they cannot grow any faster.
- $ACN(K) \leq \gamma \cdot (\frac{L}{R})(\kappa(K))$
  Here we believe a better theorem exists, with one or both of the exponents of $\frac{L}{R}$ or $\kappa(K)$ less than 1; perhaps $ACN(K) \leq \tilde{\gamma} \cdot (\frac{L}{R})(\kappa(K)^{1/2})$?
- If we know a spatial packing regime for K (see Section 2.3.2) then we can specify the ACN bound more closely:
  $ACN(K) \leq \delta \cdot (\frac{L}{R})(W)$ where W is
  - constant if packing is linear
  - $ln(\frac{L}{R})$ if packing laminar
  - $(\frac{L}{R})^{1/3}$ if packing is globular

The detailed proofs of these theorems are given in [**BS99, BS07b**]. In the next section, we give try to give an heuristic explanation, in particular, how the three packing regimes give different bounds for ACN.

7. Heuristic proofs for packing regimes and ACN

In this section we give an heuristic/intuitive analysis of how the packing in the three "regimes" influences the average crossing number.

We have a filament K of length L, think of L as very large. Assume that the spatial packing is roughly homogeneous in the sense that around each point x of

K, for each radius r, the amount of length contained in a ball of radius r about x is on the same order. More precisely, let $\ell(x,r)$ be the total length of K that lies at distance r from x. In discrete terms, the number $\ell(x,r)$ can also be viewed as the total length of K that is contained in a spherical shell of some thickness Δr about x.

The Gauss integral that gives the average crossing number can be separated into a "local" contribution proportional to total length L, due to points that are close together in arclength along the curve, and a higher-order generic contribution. We will focus on the latter, and {bound, estimate} the ACN by the simple integral

$$ACN(K) \approx \int_{x\in K} \int_{r=1}^{\infty} \frac{\ell(x,r)}{r^2} .$$

As we are assuming the filament distribution is spatially homogeneous, the function $\ell(x,r)$ is independent of x and each point $x \in K$ sees the same inner integral. So we can write

$$ACN(K) \approx L \times \int_{r=1}^{\infty} \frac{\ell(x,r)}{r^2} .$$

It is important to note that he upper limit, ∞, is larger than needed: Once radius r is large enough, call the value R (yet to be specified), that a ball about a "typical" point $x \in K$ engulfs all of K, the numerator $\ell(x,r) = 0$. So let us write

$$ACN(K) \approx L \times \int_{r=1}^{R} \frac{\ell(x,r)}{r^2} .$$

We now identify three special packing regimes: In each case, we will assume a certain scaling of $\ell(r)$, infer a consequent R, and get a particular {bound, estimate} for ACN.

Linear: The filament is strung out in space, thinking of e.g. a long string of trefoil knots tied end-to-end wandering off. Here $\ell(x,r)$ is constant, independent of r (i.e. $\ell(x,r) \approx \Delta r$). Ignoring the constant, we will write $\ell(x,r) \approx 1$. In this situation, we have to let R be as big as L in order to make sure that the whole filament K is included. So our {bound, estimate} in the *linear regime* is

$$ACN(K) \approx L \times \int_{r=1}^{L} \frac{1}{r^2} \approx L .$$

Laminar: The filament is distributed in space in a fractal-looking way, or so it fills a virtual surface. Here $\ell(x,r)$ is enough length to form a circle of radius r (thinking of this circle as the boundary of a disk of radius r about x), so $\ell(x,r)$ is proportional to r. Again ignoring the constants, we write $\ell(x,r) \approx r$. In this situation, we have on average $R \approx \sqrt{L}$. So our {bound, estimate} is

$$ACN(K) \approx L \times \int_{r=1}^{\sqrt{L}} \frac{r}{r^2} \approx L \times \log(\sqrt{L}) \approx L \log(L) .$$

Globular: The filament is distributed in space so that it essentially fills a volume. Here $\ell(x,r) \approx r^2$ and $R \approx L^{1/3}$. So

$$ACN(K) \approx L \times \int_{r=1}^{L^{1/3}} \frac{r^2}{r^2} \approx L \times L^{1/3} = L^{4/3} .$$

8. Additional comments on using geometric properties to distinguish between different "random" tangles

In Section 2.3.3 we discussed using geometric properties to distinguish two different ways of generating "random" (polygonal) filaments (see Figure 9). In this section, we add some further observations.

Recall that one polygon was produces by jiggling (many times) a 100-edge regular planar polygon using "crankshaft" moves. This is a process that has become standard for simulating random polymers. Note that this process preserves the number of edges and the lengths of all the edges, so the result is an equilateral polygon whose length is the same as the original.

The second polygon started with a regular planar 100-gon, but here we moved each vertex a random amount (chosen with uniform distribution from $[0, D]$) in a random direction, where D is a fixed large number. This produces a very zig-zag looking figure that we do not claim is a realistic polymer model. To keep the figure visible, we need to rescale it to some convenient size.

As noted earlier, one distinction between the polygons is their average exterior turning angle, which is 90° for the crankshaft randomized polygon and 120° for the other. So the total curvature is much larger for the second polygon than for the first; this is one clear and easily measured difference between the two processes, and one where rescaling has no effect.

What about compaction? We can get a rough estimate without doing an elaborate probabilistic calculation.

The process of rescaling (which we do to the second polygon) would of course change the length. To get a (very rough here) scale invariant measure of compaction, we enclose each object in a tight rectangular box with sides $2x, 2y, 2z$, calculate a mean radius, call it $\langle\rho\rangle$ as as $\sqrt{x^2+y^2+z^2}$, then take the ratio $L/\langle\rho\rangle$.

We get the following data.

100 segments	100 segments
$L = 184$	$L = 183$
$\frac{L}{\langle\rho\rangle} \approx 16$	$\frac{L}{\langle\rho\rangle} \approx 57$
ACN ≈ 42	ACN ≈ 702
$\kappa(K) \approx 100 \times \pi/2$	$\kappa(K) \approx 100 \times 2\pi/3$

We have two methods for generating "random" polygons, clearly distinguished by geometric properties that we have been discussing: compaction, average crossing number of different orders of magnitude, and total curvature.

9. Does ACN really measure "tangling"?

In Section 5 we asked whether average crossing number really measures tangling. To some extent, this question cannot be answered since "tangling" is not well defined. However, we have conducted computer experiments [**SST**] that seems to validate the use of ACN as an average measure over a statistical ensemble of filaments.

We started with 21-segment equilateral polygons and randomized them using the "crankshaft" process. The number of edges was arbitrary, but chosen small enough that nontrivial knots would be rare. We then relaxed each randomized polygon in KnotPlot, using the "minimum-distance" energy and a simple "twitch randomly and see if the energy goes down" algorithm for relaxation. We counted how many cycles of the algorithm it took to get the polygon below a threshold where we know only unknots can occur. (This is a delicate point in such an experiment: if one tries to relax too far, then the process slows down; it takes many random perturbations to move a low-energy conformation into a slightly lower one; and this would blur the differences in relaxation time that we are trying to detect.) The chart below (Figure 23) shows a fairly strong linear correlation between the ACN of the starting polygon and the amount of "time" it takes to untangle it.

We need to conduct more experiments along these lines to have a fully convincing argument, but this is a beginning of an experimental proof that ACN does effectively (albeit indirectly) measure tangling in random filament systems.

We might also note that the lower set of data in the chart shows the correlation between our other proposed measure, ABN, and relaxation. The linear correlation is stronger ($R^2 = 0.75$ vs 0.63, which is not an artifact of one set of numbers just being smaller than the other; the correlation is unchanged by rescaling; but the difference may be due to a very small number of outliers where ACN is high but ABN is low), suggesting that ABN may be an even better way to measure "tangling": we are continuing this work.

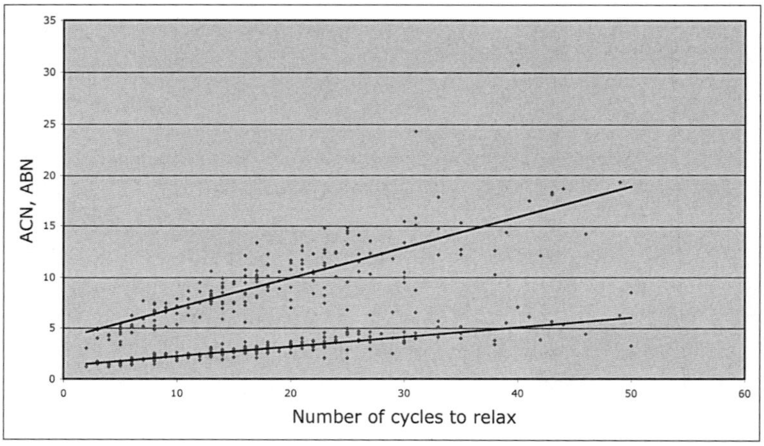

FIGURE 23. Linear correlations between "relaxation time" and ACN (or ABN) for random polygonal unknots

References

[Art97] Gustavo A. Arteca, *Self-similarity in entanglement complexity along the backbones of compact proteins*, Phys. Rev. E **56** (1997), no. 4, 4516–4520.

[AT00] G. Arteca and O. Tapia, *Relative measure of geometrical entanglement to study folding-unfolding transitions. application to analysis of molecular dynamics trajectories of proteins* in vacuo, International Journal of Quantum Chemistry **80** (2000), no. 4-5, 848–855.

[BF07a] D. Buck and E. Flapan, *A model of dna knotting and linking*, Knot Theory for Scientific Objects, A. Kawauchi, Ed., OCAMI Studies, vol. 1, Osaka Municipal Universities Press, Osaka, 2007.

[BF07b] Dorothy Buck and Erica Flapan, *A topological characterization of knots and links arising from site-specific recombination*, J. Phys. A **40** (2007), no. 41, 12377–12395. MR2394909

[BPR05] Justyna Baranska, Piotr Pieranski, and Eric J. Rawdon, *Ropelength of tight polygonal knots*, Physical and numerical models in knot theory, Ser. Knots Everything, vol. 36, World Sci. Publ., Singapore, 2005, pp. 293–321. MR2197946 (2007j:74076)

[BS99] Gregory Buck and Jonathan Simon, *Thickness and crossing number of knots*, Topology Appl. **91** (1999), no. 3, 245–257. MR1666650 (99k:57026)

[BS07a] G. Buck and J. Simon, *Models of entanglement*, Knot Theory for Scientific Objects, A. Kawauchi, Ed., OCAMI Studies, vol. 1, Osaka Municipal Universities Press, Osaka, 2007.

[BS07b] Gregory R. Buck and Jonathan K. Simon, *Total curvature and packing of knots*, Topology Appl. **154** (2007), no. 1, 192–204. MR2271783 (2007i:58011)

[BSSS08] Gregory Buck, Robert G. Scharein, Jeffrey Schnick, and Jonathan Simon, *Accessibility and occlusion of biopolymers, ray tracing of radiating tubes, and the temperature of a tangle*, Physical Review E (Statistical, Nonlinear, and Soft Matter Physics) **77** (2008), no. 1, 011803.

[Buc] Gregory R. Buck, Using the "analog computer" method of careful measurement with actual rope, G. Buck noted around 1995-6 that the minimum ropelength of a trefoil knot is approximately 32.5.

[Buc98] Gregory Buck, *Four-thirds power law for knots and links*, Nature **392** (1998), no. 6673, 238–239.

[CDG01] Jason Cantarella, Dennis DeTurck, and Herman Gluck, *Upper bounds for the writhing of knots and the helicity of vector fields*, Knots, braids, and mapping class groups—papers dedicated to Joan S. Birman (New York, 1998), AMS/IP Stud. Adv. Math., vol. 24, Amer. Math. Soc., Providence, RI, 2001, pp. 1–21. MR1873104 (2003j:58018)

[Cha64] G. D. Chakerian, *On some geometric inequalities*, Proc. Amer. Math. Soc. **15** (1964), 886–888. MR0170273 (30 #512)

[CKS98] J. Cantarella, R. Kusner, and J. Sullivan, *Tight knot values deviate from linear relations*, Nature **392** (1998), 237–238.

[CLMS05] M. Carlen, B. Laurie, J. H. Maddocks, and J. Smutny, *Biarcs, global radius of curvature, and the computation of ideal knot shapes*, Physical and numerical models in knot theory, Ser. Knots Everything, vol. 36, World Sci. Publ., Singapore, 2005, pp. 75–108. MR2197935 (2006k:58015)

[DDK+03] Y. Diao, A. Dobay, R. B. Kusner, K. Millett, and A. Stasiak, *The average crossing number of equilateral random polygons*, J. Phys. A **36** (2003), no. 46, 11561–11574. MR2025860 (2005a:82039)

[DE02] Yuanan Diao and Claus Ernst, *The crossing numbers of thick knots and links*, Physical knots: knotting, linking, and folding geometric objects in $\mathbb{R}^3$ (Las Vegas, NV, 2001), Contemp. Math., vol. 304, Amer. Math. Soc., Providence, RI, 2002, pp. 163–174. MR1953338 (2003m:57010)

[DE05] ______, *The average crossing number of Gaussian random walks and polygons*, Physical and numerical models in knot theory, Ser. Knots Everything, vol. 36, World Sci. Publ., Singapore, 2005, pp. 275–292. MR2197945 (2006j:82020)

[DH08] P.M. Diesinger and D.W. Heermann, *Average crossing number of gaussian and equilateral chains with and without excluded volume*, The European Physical Journal B - Condensed Matter and Complex Systems **62** (2008), no. 2, 209–214.

[Dia93] Yuanan Diao, *Minimal knotted polygons on the cubic lattice*, J. Knot Theory Ramifications **2** (1993), no. 4, 413–425. MR1247576 (94j:57005)

[DSS08] I. K. Darcy, R. G. Scharein, and A. Stasiak, *3D visualization software to analyze topological outcomes of topoisomerase reactions*, Nucl. Acids Res. (2008), gkn192.

[EEA00] T. Edvinsson, C. Elvingson, and G. Arteca, *Variations in molecular compactness and chain entanglement during the compression of grafted polymers*, Macromolecular Theory and Simulations **9** (2000), no. 7, 398–406.

[Fár49] István Fáry, *Sur la courbure totale d'une courbe gauche faisant un nœud*, Bull. Soc. Math. France **77** (1949), 128–138. MR0033118 (11,393h)

[Fár50] ______, *Sur certaines inégalites géométriques*, Acta Sci. Math. Szeged **12** (1950), no. Leopoldo Fejer et Frederico Riesz LXX annos natis dedicatus, Pars A, 117–124. MR0038090 (12,353b)

[Fen51] Werner Fenchel, *On the differential geometry of closed space curves*, Bull. Amer. Math. Soc. **57** (1951), 44–54. MR0040040 (12,634d)

[FHW94] Michael H. Freedman, Zheng-Xu He, and Zhenghan Wang, *Möbius energy of knots and unknots*, Ann. of Math. (2) **139** (1994), no. 1, 1–50. MR1259363 (94j:58038)

[Gho07] Mohammad Ghomi, *h-principles for curves and knots of constant curvature*, Geom. Ded. **127** (2007), no. 1, 19–35, (earlier preprint 2005).

[GM99] Oscar Gonzalez and John H. Maddocks, *Global curvature, thickness, and the ideal shapes of knots*, Proc. Natl. Acad. Sci. USA **96** (1999), no. 9, 4769–4773 (electronic). MR1692638 (2000c:57008)

[Gra01] P. Grassberger, *Opacity and entanglement of polymer chains*, Journal of Physics A: Mathematical and General **34** (2001), 9959–9963.

[Hop35] Heinz Hopf, *Über die Drehung der Tangenten und Sehnen ebener Kurven*, Compositio Math. **2** (1935), 50–62. MR1556906

[JvRSWW92] E. J. Janse van Rensburg, D. A. W. Sumners, E. Wasserman, and S. G. Whittington, *Entanglement complexity of self-avoiding walks*, J. Phys. A **25** (1992), no. 24, 6557–6566. MR1210877 (94d:82043)

[KBM+96] Vsevolod Katritch, Jan Bednar, Didier Michoud, Robert G. Scharein, Jacques Dubochet, and Andrzej Stasiak, *Geometry and physics of knots*, Nature **384** (1996), no. 6605, 142–145.

[LSDR99] R. A. Litherland, J. Simon, O. Durumeric, and E. Rawdon, *Thickness of knots*, Topology Appl. **91** (1999), no. 3, 233–244. MR1666649 (99k:57025)

[LZC02] Jie Liang, Jinfeng Zhang, and Rong Chen, *Statistical geometry of packing defects of lattice chain polymer from enumeration and sequential monte carlo method*, The Journal of Chemical Physics **117** (2002), no. 7, 3511–3521.

[MG07] J. McAtee Ganatra, *Knots of constant curvature*, J. Knot Theory Ramifications **16** (2007), no. 4, 461–470, accepted 2005, preprint 2004 arXiv:math/0403089v1.

[Mil50] J. W. Milnor, *On the total curvature of knots*, Ann. of Math. (2) **52** (1950), 248–257. MR0037509 (12,273c)

[MR03] Kenneth C. Millett and Eric J. Rawdon, *Energy, ropelength, and other physical aspects of equilateral knots*, J. Comput. Phys. **186** (2003), no. 2, 426–456.

[Neg91] Seiya Negami, *Ramsey theorems for knots, links and spatial graphs*, Trans. Amer. Math. Soc. **324** (1991), no. 2, 527–541. MR1069741 (92h:57014)

[Pie98] Piotr Pierański, *In search of ideal knots*, Ideal knots, Ser. Knots Everything, vol. 19, World Sci. Publ., River Edge, NJ, 1998, pp. 20–41. MR1702021

[PP01] P. Pieranski and S. Przybyl, *Ideal trefoil knot*, Phys. Rev. E (3) **64** (2001), no. 3, part 1, 031801, 5. MR1856439 (2002h:58031)

[PP02] Piotr Pieranski and Sylwester Przybyl, *In search of the ideal trefoil knot*, Physical knots: knotting, linking, and folding geometric objects in $\mathbb{R}^3$ (Las Vegas, NV, 2001), Contemp. Math., vol. 304, Amer. Math. Soc., Providence, RI, 2002, pp. 153–162. MR1953337 (2003m:57022)

[PPS01] P. Pieranski, S. Przybyl, and A. Stasiak, *Gordian unknots*, 2001, Preprint arXiv.org:physics/0103080.

[Ran94a] Richard Randell, *An elementary invariant of knots*, Random knotting and linking (Vancouver, BC, 1993), Ser. Knots Everything, vol. 7, World Sci. Publ., River Edge, NJ, 1994, pp. 47–54. MR1474262 (98e:57015)

[Ran94b] ———, *An elementary invariant of knots*, J. Knot Theory Ramifications **3** (1994), no. 3, 279–286, Random knotting and linking (Vancouver, BC, 1993). MR1291860 (95e:57019)

[RAT02] C.T. Reimann, G. Arteca, and O. Tapia, *Proteins* in vacuo. *a connection between mean overcrossing number and orientationally-averaged collision cross section*, Phys. Chem. Chem. Phys. **4** (2002), 4058–4064.

[Raw98] Eric J. Rawdon, *Approximating the thickness of a knot*, Ideal knots, Ser. Knots Everything, vol. 19, World Sci. Publ., River Edge, NJ, 1998, pp. 143–150. MR1702029

[Raw00] ———, *Approximating smooth thickness*, J. Knot Theory Ramifications **9** (2000), no. 1, 113–145. MR1749504 (2001a:57017)

[RB03] Peter Røgen and Henrik Bohr, *A new family of global protein shape descriptors*, Math. Biosci. **182** (2003), no. 2, 167–181. MR1965354 (2004a:92048)

[RF03] Peter Røgen and Boris Fain, *Automatic classification of protein structure by using Gauss integrals*, Proceedings of the National Academy of Sciences **100** (2003), no. 1, 119–124.

[RS02] Eric J. Rawdon and Jonathan Simon, *Möbius energy of thick knots*, Topology Appl. **125** (2002), no. 1, 97–109. MR1931178 (2004c:58029)

[RST] R. Randell, J. Simon, and J. Tokle, *Möbius transformations of polygons and partitions of 3-space*, Preprint arXiv:math/0602466.

[Sch54] Horst Schubert, *Über eine numerische Knoteninvariante*, Math. Z. **61** (1954), 245–288. MR0072483 (17,292a)

[Sch85] Martin G. Scharlemann, *Unknotting number one knots are prime*, Invent. Math. **82** (1985), no. 1, 37–55. MR808108 (86m:57010)

[SST] R. Scharein, J. Simon, and M. Tucker, *Average bridging number for knots and other filament systems*, In preparation. Includes material presented at AMS meetings 10/06 and 11/07 and BIRS-Banff symposium 5/07.

[Sum95] De Witt Sumners, *Lifting the curtain: using topology to probe the hidden action of enzymes*, Notices Amer. Math. Soc. **42** (1995), no. 5, 528–537. MR1324733 (96a:92006)

[Sum07] D. Sumners, *Dna topology: Experiments and analysis*, Knot Theory for Scientific Objects, A. Kawauchi, Ed., OCAMI Studies, vol. 1, Osaka Municipal Universities Press, Osaka, 2007.

[Tab03] Serge Tabachnikov, *The tale of a geometric inequality*, MASS selecta, Amer. Math. Soc., Providence, RI, 2003, pp. 257–262. MR2027183

[Var] VariousAuthors, see e.g. papers or online postings by Colin Adams & students, J. Calvo, Y. Diao, Y. Huh, E.J. Janse van Rensburg, C. McCabe, M. Meissen, S. Negami, S. Oh, R. Randell, E. Rawdon & R. Scharein, R. Trapp & students. Search online for "knots" together with any one of "edge number", "stick number", "lattice edge number", "equilateral edge number" of knots.

[VCL+98] A. Vologodskii, N. Crisona, B. Laurie, P. Pieranski, V. Katritch, J. Dubochet, and A. Stasiak, *Sedimentation and electrophoretic migration of dna knots and catenanes*, J. Mol. Biol. **278** (1998), no. 1, 1–3.

[Whi37] Hassler Whitney, *On regular closed curves in the plane*, Compositio Math. **4** (1937), 276–284. MR1556973

Department of Mathematics, University of Iowa, Iowa City, IA 52242
E-mail address: `jsimon@math.uiowa.edu`

Index

Titles in This Series

66 **Dorothy Buck and Erica Flapan, Editors,** Applications of knot theory (San Diego, California, January 2008)

65 **L. L. Bonilla, A. Carpio, J. M. Vega, and S. Venakides, Editors,** Recent advances in nonlinear partial differential equations and applications (Toledo, Spain, June 2006)

64 **Reinhard C. Laubenbacher, Editor,** Modeling and simulation of biological networks (San Antonio, Texas, January 2006)

63 **Gestur Ólafsson and Eric Todd Quinto, Editors,** The radon transform, inverse problems, and tomography (Atlanta, Georgia, January 2005)

62 **Paul Garrett and Daniel Lieman, Editors,** Public-key cryptography (Baltimore, Maryland, January 2003)

61 **Serkan Hoşten, Jon Lee, and Rekha R. Thomas, Editors,** Trends in optimization (Phoenix, Arizona, January 2004)

60 **Susan G. Williams, Editor,** Symbolic dynamics and its applications (San Diego, California, January 2002)

59 **James Sneyd, Editor,** An introduction to mathematical modeling in physiology, cell biology, and immunology (New Orleans, Louisiana, January 2001)

58 **Samuel J. Lomonaco, Jr., Editor,** Quantum computation: A grand mathematical challenge for the twenty-first century and the millennium (Washington, DC, January 2000)

57 **David C. Heath and Glen Swindle, Editors,** Introduction to mathematical finance (San Diego, California, January 1997)

56 **Jane Cronin and Robert E. O'Malley, Jr., Editors,** Analyzing multiscale phenomena using singular perturbation methods (Baltimore, Maryland, January 1998)

55 **Frederick Hoffman, Editor,** Mathematical aspects of artificial intelligence (Orlando, Florida, January 1996)

54 **Renato Spigler and Stephanos Venakides, Editors,** Recent advances in partial differential equations (Venice, Italy, June 1996)

53 **David A. Cox and Bernd Sturmfels, Editors,** Applications of computational algebraic geometry (San Diego, California, January 1997)

52 **V. Mandrekar and P. R. Masani, Editors,** Proceedings of the Norbert Wiener Centenary Congress, 1994 (East Lansing, Michigan, 1994)

51 **Louis H. Kauffman, Editor,** The interface of knots and physics (San Francisco, California, January 1995)

50 **Robert Calderbank, Editor,** Different aspects of coding theory (San Francisco, California, January 1995)

49 **Robert L. Devaney, Editor,** Complex dynamical systems: The mathematics behind the Mandlebrot and Julia sets (Cincinnati, Ohio, January 1994)

48 **Walter Gautschi, Editor,** Mathematics of Computation 1943–1993: A half century of computational mathematics (Vancouver, British Columbia, August 1993)

47 **Ingrid Daubechies, Editor,** Different perspectives on wavelets (San Antonio, Texas, January 1993)

46 **Stefan A. Burr, Editor,** The unreasonable effectiveness of number theory (Orono, Maine, August 1991)

45 **De Witt L. Sumners, Editor,** New scientific applications of geometry and topology (Baltimore, Maryland, January 1992)

44 **Béla Bollobás, Editor,** Probabilistic combinatorics and its applications (San Francisco, California, January 1991)

43 **Richard K. Guy, Editor,** Combinatorial games (Columbus, Ohio, August 1990)

42 **C. Pomerance, Editor,** Cryptology and computational number theory (Boulder, Colorado, August 1989)

41 **R. W. Brockett, Editor,** Robotics (Louisville, Kentucky, January 1990)

ISBN 978-0-8218-4466-3

9 780821 844663

PSAPM/66